Małgorzata Dominika Brzeska

RF Modelling and Characterization of Tyre Pressure Sensors and Vehicle Access Systems

Karlsruher Forschungsberichte
aus dem Institut für Hochfrequenztechnik und Elektronik

Herausgeber: Prof. Dr.-Ing. Thomas Zwick

Band 78

RF Modelling and Characterization of Tyre Pressure Sensors and Vehicle Access Systems

by
Małgorzata Dominika Brzeska

KIT Scientific Publishing

Dissertation, Karlsruher Institut für Technologie (KIT)
Fakultät für Elektrotechnik und Informationstechnik, 2014

Impressum

KIT Scientific Publishing

Karlsruher Institut für Technologie (KIT)
KIT Scientific Publishing
Straße am Forum 2
D-76131 Karlsruhe

KIT Scientific Publishing is a registered trademark of Karlsruhe Institute of Technology. Reprint using the book cover is not allowed.

www.ksp.kit.edu

Print on Demand 2015

ISSN 1868-4696
ISBN 978-3-7315-0348-4
DOI 10.5445/KSP/1000045931

Vorwort

Wir werden fast täglich überrascht von Ideen für die Autos der Zukunft: sie werden smart und kontrollieren und steuern viele relevante Funktionen selbständig und automatisch. Die Arbeit von Frau Brzeska fügt sich in diese Entwicklung ideal ein. Sie entwirft und optimiert ein Reifendruck-Messsystem zur automatischen und kontinuierlichen Messung des Druckzustandes in den Reifen. Drei Schwerpunkte ergeben sich hieraus für die Arbeit: Messung des Reifendrucks, Übertragung der Information drahtlos aus dem Reifen und simultanes Empfangen der gesendeten Informationen aller Reifen zur zentralen Kontrolle und Verwertung.

Aus diesen Aufgaben resultieren die neuen Beiträge der Arbeit von Frau Brzeska: Integration des Drucksensors mit dem Sender zu einem Modul mit hoher mechanischer Stabilität und langer Lebensdauer; Bestimmung der Übertragungseigenschaften vom Inneren des Reifens nach außen für alle Reifen unabhängig vom jeweiligen Material des Gummis und des integrierten Stahlmantels; Optimierung des Empfängerposition zur Sicherung des Empfangs der Reifendruckinformation auch bei schwierigen Bedingungen wie Schmutz oder Nässe.

Frau Brzeska hat diese mit hohem Forschungsaufwand verbundenen Aufgaben und das Designs des Reifendruck-Messsystem hervorragend gelöst. Ihre Ergebnisse werden in die Reifendruck-Messsysteme der Zukunft einfließen.

Ich wünsche Frau Brzeska auf Ihrem weiteren Weg, dass ihre wissenschaftliche Kreativität und ihr Ideenreichtum sie zu technischen und wirtschaftlichen Erfolgen führt.

Prof. Dr.-Ing. Dr. h.c. Dr.-ing. E.h. mult. Werner Wiesbeck

RF Modelling and Characterization of Tyre Pressure Sensors and Vehicle Access Systems

Zur Erlangung des akademischen Grades eines

DOKTOR-INGENIEURS

der Fakultät für
Elektrotechnik und Informationstechnik
Karlsruher Institut für Technologie

genehmigte

DISSERTATION

von

Dipl.-Ing. Małgorzata Dominika Brzeska
aus Gdańsk

Tag der mündlichen Prüfung:	26.02.2014
Hauptreferent:	Prof. Dr.-Ing. Dr.h.c. Dr.-Ing. E.h. Werner Wiesbeck
Korreferent:	Univ. Prof. Ing. Dipl.-Ing. Dr.-Ing. Christoph Mecklenbräuker

Danksagung

Die vorliegende Dissertation entstand im Rahmen meiner Tätigkeit in Kooperation zwischen der Universität Karlsruhe, Institut für Höchstfrequenztechnik und Elektronik (IHE) und der Firma Siemens VDO, später Continental AG in Regensburg. Mein besonderer Dank gilt allen Personen, die diese Zusammenarbeit ermöglicht haben: Herrn Prof. Dr.-Ing. Dr.h.c. Dr.-Ing. E.h. Werner Wiesbeck und der gesamten Abteilung Body und Chassis Elektronik, besonders Herrn Dr. Guy Aymar Chakam, Herrn Herbert Froitzheim und Herrn Christian Schneider.

Herrn Prof. Dr.-Ing. Dr.h.c. Dr.-Ing. E.h. Werner Wiesbeck danke ich für die fachliche Unterstützung und die vielen Diskussionen über die Arbeit. Neben dieser wertvollen fachlichen Unterstützung sind die vermittelten Werte und der Arbeitsstil wichtige Aspekte, die für mich über die Arbeit hinaus von großer Bedeutung bleiben und mir Impulse für die persönliche Weiterentwicklung gegeben haben.

Herrn Univ. Prof. Ing. Dipl.-Ing. Dr.-Ing. Christoph Mecklenbräuker danke ich sehr für die übernahme des Korreferats, das große Interesse an dieser Arbeit, für alle Anregungen und freundliche Aufnahme in Wien als auch für die interessanten Diskussionen.

Herrn Prof. Dr.-Ing. Thomas Zwick danke ich für die kurze, wertvolle Arbeitsdiskussion und Unterstützung.

Allen Kollegen, mit denen ich im Laufe meiner Tätigkeit zusammengearbeitet habe, möchte ich meinen Dank für ein hervorragendes Arbeitsklima aussprechen. Ich danke besonders Suhas Jawale, Dr. Martin Weinberger, Mirko Todorovic, Niels Koch, Arturo Arrango, Peter Turban und Alois Winzinger. Alexis Morin danke ich für die Zusammenarbeit während meines Aufenthaltes in Toulouse.

Bei Herrn Dr. Grzegorz Adamiuk möchte ich mich für sein gewissenhaftes Korrekturlesen und die konstruktive Kritik dieser Arbeit bedanken. Frau Dr. Małgorzata Janson und Herrn Dr. Mario Pauli danke ich für die wertvolle Diskussion zu der finalen Präsentation.

Frau Karin Richter und allen Institutskollegen danke ich für ihre stete Unterstützung und den Informationsaustausch.

Die Möglichkeit, Studenten auf dem Weg zur Abschlussarbeit zu unterstützen, hat mir sehr viel Freude bereitet. Ich danke für die hervorragende Arbeit und den stets guten Humor Herren Paul Tutu, Tudor Ziemann, Yannick Dieudonne und Frau Dessislava Dimitrova.

Mein großer Dank gilt meiner Familie besonders meiner Mutter
- Dziekuję Ci za wszystko Mamo!
meinem Bruder Piotr Domaradzki und meinem Freund und Partner Andreas Schäfer.

Contents

Acronyms and Symbols

List of Constants

c_0	speed of light $2.997925 \times 10^8\ m/s$ in vacuum
ε_0	permittivity of vacuum $8.854 \times 10^{-12} As/Vm$
μ_0	permeability $4\pi \times 10^{-7}$ of vacuum

Mathematical Notations and Symols

j	imaginary unit $j = \sqrt{-1}$
sin	sinus
$=$	equal
$+$	plus
$-$	minus
x	the vector cross product of two vectors
$\%$	percent

Acronyms

1D	One-Dimensional
2D	Two-Dimensional
3D	Three-Dimensional
ABS	Anti-lock Braking System
AC	Access Card
acc.	according
AE	American English
AG	Aktiengesellschaft, engl. incorporated company
AM	Amplitude Modulation
ATMEL	Atmel corporation
avg.	Average
BEIDU	chineese satellite navigation system
C2C	Car to Car
C2X	Car to X
CAD	Computer Aided Design
CEM	Computational Electromagnetics
CU	Control Unit
DEKRA	Deutscher Kraftfahrzeug-Überwachungs-Verein
DOT	Department of Transportation
DUT	Device Under Test

DVB-T	Digital Video Broadcasting - Terrestrial
ECE	Economic Commission for Europe
ECU	Electronic Control Unit
EM	Electromagnetic
EMSS	Electromagnetic Software and Systems
Eq.	Equation
ESC	Electronic Stability Control
FDTD	Finite Difference Time Domain
FEKO	Feldberechnung für Kïoper mit beliebiger Oberfläche, engl. Field calculation for bodies with arbitrary surface
FEM	Finite Element Method
FM	Frequency Modulation
GLONASS	Global Navigation Satellite System
GPS	Global Positioning System
GTEM	Gigahertz Transverse ElectroMagnetic
HF	High Frequency
HFSS	High Frequency Solution Software
IC	Identification Card
ID	Identification Device
IFA	Inverted F Antenna
ILA	Inverted L Antenna
ISM	Industrial Scientific Medical
ITPMS	Indirect Tyre Pressure Monitoring System
LED	Light Emitting Diode
LF	Low Frequency
MATLAB	MATrix LABoratory
max.	maximum
meas.	measurement
min.	minimum
MoM	Method of Moments
MUT	Material Under Test
NAFTA	North American Free Trade Agreement
NHTSA	National Highway Traffic Safety Administration
NIST	National Institute of Standards and Technology
NWA	Network Analyser
OEM	Original Equipment Manufacturer
P.V.C	Polyvinylchlorid

PA66	Polyamid type 66
PASE	Passive Start and Entry
PC	Personal Computer
PCB	Printed Circuit Board
PEC	Perfect Electric conductor
PIFA	Planar Inverted F Antenna
PSPICE	Personal Simulation Program with Integrated Circuit Emphasis
PTFE	Polytetrafluoroethylene
RF	Radio Frequency
RKE	Remote Keyless Entry
RMS	Root Mean Square
RoF	Run on Flat
RoW	Rest of the World
SDARS	Satellite Digital Audio Radio Service
SEP	Surface Equivalence Principle
sim.	simulation
SSR	Self Supporting Runflat
TMC	Traffic Message Channel
TPMS	Tyre Pressure Monitoring System
TWI	Tread Wear Indicator
UI	User Interface
US	United States
UWB	Ultra Wide Band
VAS	Vehicle Access System
VDO	consolidated DEUTA (Deutsche Tachometerwerke GmbH) OTA (OTA Apparate GmbH))
VEP	Volume Equivalence Principle
VNWA	Vector Network Analyser
WE	Western Europe
WU	Wheel Unit

Greek Symbols

λ	Wavelength in m/s
θ	Elevation angle in spherical coordinate system
φ	Azimuth angle in spherical coordinate system
η	Efficiency

Latin Symbols

E	Electrical Field Strength
E_x	x component of field E

E_y	y component of field E
E_z	z component of field E
E_t	total field E
G	Gain
G_{WU}	Wheel Unit Gain
D	Directivity
d	Distance
C	Radiation pattern
x, y, z	Coordinates in cartesian coordinate system
f	frequency
f_{PIFA}	resonant frequency of PIFA
P_{rad}	Radiated power
P_{in}	Input power
$\vec{J}$	Current density
σ	Conductivity
r	Radius in spherical coordinate system
w	Width
l	Length
Z_{in}	Input Impedance
P_r	Received power
P_t	Transmitter power
R_{rad}	Radiation resistance
h_t	Height - key fob position
h_r	Height - receiver position
L	Inductivity

Units

dB	Decibel
dBi	dB relative to an isotropic radiator
$^\circ$	Degree
$^\circ$C	Celsius degree
m	Metre
mm	Millimetre
cm	Centimentre
MHz	Megahertz
kHz	Kilohertz
Hz	Hertz

1 Introduction

Core topics of the underlying work are the vehicle access systems such as PAssive Start and Entry (PASE) and Remote Keyless Entry (RKE) as well as Tyre Pressure Monitoring System (TPMS).

The focus is on the two following goals. The first is the development of antennas and functionality analysis of the mentioned systems from RF (Radio Frequency) point of view. It is accompanied by an analysis and improvement of system parts. The requirements for the performance of the above mentioned systems are growing. These result in higher demands on antennas. Though proper functionality of the systems relies upon reliable communication between the transmitter and the receiver, antennas are very often an overlooked component in vehicle systems. As also aesthetic and physical aspects call for innovative antenna design, developed to address specific requirements, simple wire antennas as monopoles are not anymore sufficient to support future systems. Therefore antennas become a very important part in modern vehicle development. Crucial is the examination and the development of the Wheel Unit antennas for TPMS as well as of the receiver for both TMPS and RKE. Also the analysis of the electrical properties of the tyres and the classification of the on-market available tyres is of high importance. These investigations form the fundamentals for the second goal, which is the modeling of both systems as well as utilization of the simulation processes. In addition to diverse antenna models and an appropriate vehicle model, also a novel, scalable tyre simulation model is developed within this work.

The overall objective of this work is to advance the state-of-the-art vehicular electromagnetic simulation taking into account the vehicle body and nearest surroundings.

1.1 Motivation

Motivation for extended investigation on the access systems and the tyre pressure monitoring is first of all the enhancement of the vehicle and passengers safety but also comfort improvement. The importance of the research is argued in the following paragraphs and the link to the very significant RF thread for the systems is shown.

The development of the vehicle to the stage as it is well known now (Fig. 1.1a) took place in steps. The first four-wheeler was introduced and sold by Karl Benz in 1893 (Fig. 1.1b [19]). Almost sixty years later, beginning in 1950s, the first RF application, namely AM radio broadcasting was available for personal use in cars [20].

Since then the amount of electronics being implemented in the modern car has grown tremendously over the years, it has penetrated almost every vehicle function. Today an ever increasing number of radio services has become available. Multifunction antenna systems are required. The antennas have to be assembled not only to meet the RF requirements but they have to satisfy the vehicle design and have minimum impact on the aesthetics and aerodynamics. Very often, when considering electromagnetic specification and small available space, the mechanical and electrical requirements are in opposition. Therefore a trade-off has to be achieved and it is a major challenge in design of an automotive antenna.

(a) Contemporary vehicle: Volvo S80

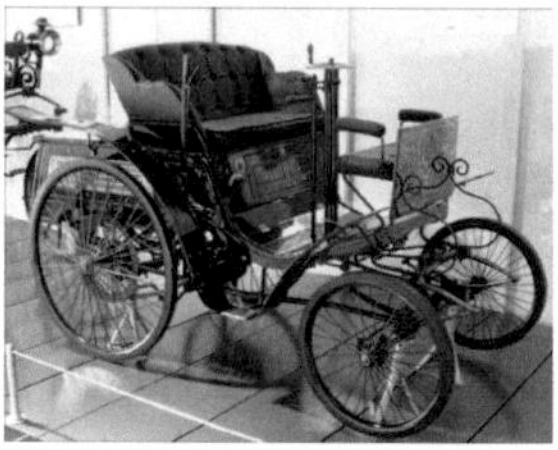

(b) Karl Benz's "Velo" model (1894)

Figure 1.1: Comparison of the first and contemporary vehicle

Meanwhile the implementation of antennas supports not only comfort but also the most important safety related functions in a vehicle. Both of the systems PASE and RKE contribute to the driver convenience in accessing the vehicle. With the help of these systems unauthorized car entrance may be omitted. One of the functions that contribute solely to the safety is the possibility to monitor the tyre pressure. In fact, various statistics show how important properly inflated tyres are. The US National Highway Traffic Safety Administration (NHTSA) estimates that 23.000 accidents and 535 fatalities per year involve flat tyres or blowouts, both of which can result from improperly inflated tyres [17]. According surveys in France showed that nine percent of all fatal road accidents are the result of under-inflated tyre. The German DEKRA (Deutscher Kraftfahrzeug-Überwachungs-Verein) found that tyre problems are linked to an estimated 41% of all injury accidents, 75% of vehicles on the road had at least one under-inflated tyre [21]. These statistics are a very strong argument and show the importance of the introduction of fully automated TPMS to future vehicles.

This was recognized and TPMS is already mandatory starting from September 2007 in all newly certified vehicles in the USA. The European Union also supports the development with various research programmes and tyre pressure sensors are expected to become standard equipment in vehicles within the next ten years. Thus, TPMS will belong to standard equipment of a vehicle in the future. For such a system, an increased confidence of operation is necessary.

Market growth is accompanied by an increased necessity to get a better understanding and to optimize the system development. Until now automobile antenna design relied only on the experience and designer intuition as well as numerous measurements. Considering the variety of different vehicle bodies, diverse tyre sizes, types and manufacturers, varied antenna solutions, the time investment and directly related cost for the system analysis is very high. In order to reduce the number of measurements together with the expenses, simulation process for system performance testing should be established used for prediction of the system behaviour. It would be advantageous to develop both a TPMS as well as access system (PASE, RKE) simulator with the help of a virtual environment for a certain number of different experimental scenarios including exchangeable vehicle bodies, tyres, rims, sensor antennas and receivers and possibly reduce the number of actual physical tests. To the author's knowledge, simulations of such complexity have not yet been conducted.

Summing up the arguments for the development of the system simulation are apart from technical advancement, also the economic and the safety reasons.

1.2 Market Situation

In order to show the importance of the vehicle access and tyre pressure monitoring systems, it is worth to analyse the automotive market situation in passenger vehicle sector.

From the origin of typical vehicle access system that involves a conventional mechanical key for locking and unlocking each door separately, the central locking technology emerged. Remote Keyless Entry (RKE) was introduced and has become a standard feature. Next generation is the Passive Start and Entry System (PASE). Each produced vehicle is equipped with one of these systems. Mechanical key exists only as a backup for radio controlled systems. Worldwide the number of new cars fitted with RKE is reaching saturation point. Fitment rates are far above 90% [22]. PASE as a more comfortable and innovative system is still an option. This means that practically all manufactured vehicles, in year 2010 over 70 millions according to Fig. 1.2 [18], are equipped with antenna based access systems.

TPMS system has been relatively lately introduced to the market in year 2002 and became already mandatory in USA. European countries and the rest of the world also show an increased interest. The statistics show that in 2010 more than 30 million vehicles had the system implemented (Fig. 1.3a) and the implementation rate is growing. This introduces an opportunity for the OEMs (Original Equipment Manufacturers) and after market suppliers. New part of market opens, where shares have still to be divided and will stabilize in the next few years. The market value reached 1.300 million Euros in 2010 (Fig. 1.3b).

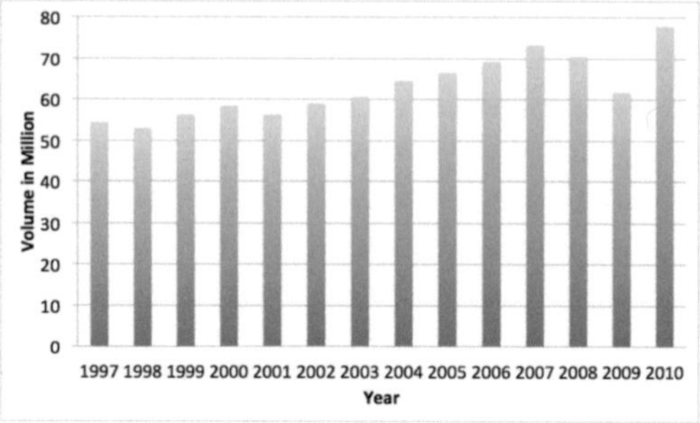

Figure 1.2: Global production of motor vehicles (passenger and commercial)

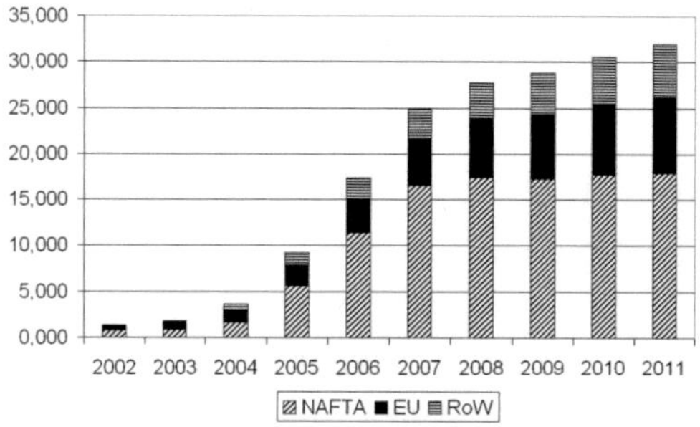

(a) Size (million systems)

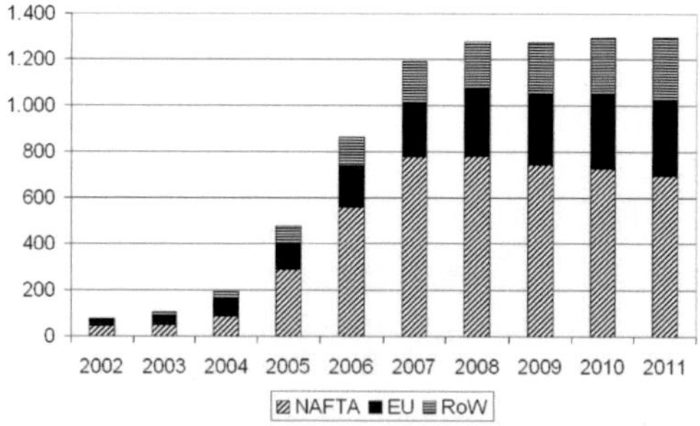

(b) Market value (million USD)

Figure 1.3: Worldwide direct TPMS market

The predictions show that within the next few years the percentage of applied systems will grow and then it will stay roughly constant year after year and will be only dependent from the number of manufactured vehicles.

1.3 Thesis Overview

Due to the complexity of the considered topics and their mutual dependence, in order to keep the thesis well structured it is divided into Chapters and belonging Appendixes. The explanation of the structure follows beneath and is additionally illustrated in Fig. 1.4 .

Following the introduction in Chapter 1, Chapter 2 gives an overview about the functionality of all three considered systems: TPMS, RKE and PASE. The architecture of the systems is introduced with special attention to RF relevant parts and the nearest surrounding. This has an essential influence on the antenna functionality. The Wheel Unit antennas positioning is prescribed by the functional assignment. For the system receiver, however different positioning scenarios are taken into account. The optimal solution for the placement is presented.

Simulation and evaluation procedure for optimal receiver antenna position within the vehicle structure is explained in Appendix C. Final placement is proposed.

Not only the receiver but each antenna design is dedicated to the environment where the antenna is deployed. Therefore also the rim and the tyre influence is analyzed. Regarding the vast variety of tyres that are available on the market, their influence has to be inspected. As it is impossible for time and cost reasons to measure each tyre, a way to classify the tyres and the final classification results are presented in Chapter 3. It is explained how the tyres are built. Mechanical classification criteria are listed. A selection of popular and most representative tyres is made for further investigation.

Additional information about the tyres, partial results and steps essential for final statements in Chapter 3 are gathered in Appendix A.

Before determining tyre influence on the antenna radiation pattern, the tyre structure is analyzed from electromagnetic point of view in Chapter 4. For each tyre region: tread and sidewalls as well as for each layer the electric material properties such as electrical permittivity and loss factor are investigated and a method to measure these is presented. The accuracy of the mentioned measurements is verified. Also embedding materials for the Wheel Units are characterized. All obtained values will be used for antenna and tyre simulation models.

Results of tyre rubber material measurements are listed in Appendix D. Chapter 4 comprehends the final result values.

In order to know what influences the tyre radiation pattern mostly, several Wheel Unit antennas are designed, measured and compared. The radiation pattern measurement setup is described in Appendix B. Also achieved characteristic values are listed in the same Appendix. Chapter 5 deals with small antenna development, both for the Wheel Units as well as for the receiver.

Chapter 6 gives an overview of the automotive electromagnetic simulation. The advantages and disadvantages of the Method of Moments for large-scale problems are discussed. An equivalent vehicle model is presented and the simulation results for PASE, RKE and TPMS are analyzed and obtained.

The schematic overview in Fig. 1.4 presents the thesis structure in a clear, well arranged way. It is depicted that Chapter 1 and Chapter 2 open the Thesis and introduce the reader the the considered topics. Chapters 3, 4 and 5 are parallel and may be considered by the reader in alternative order - the boundaries between these Chapters are not strictly confined and the reviewed topics refer to each other. In this way with accompanying Appendixes a base is formed for Chapter 6, which describes the final results.

Chapter 7 closes the Thesis with a brief summary.

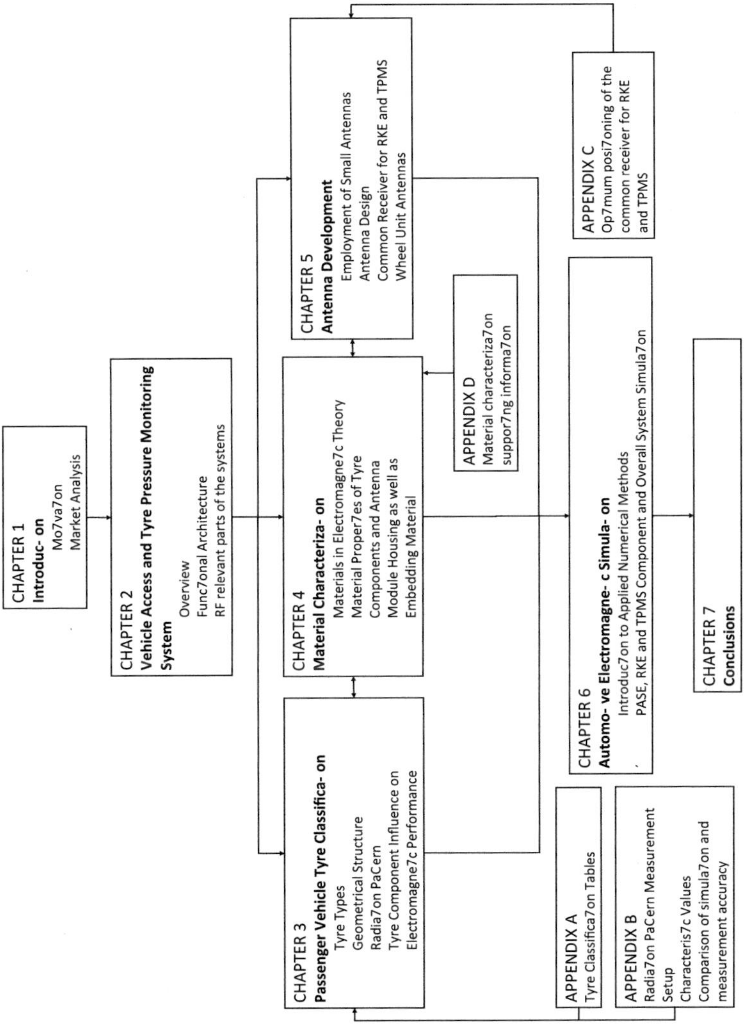

Figure 1.4: Thesis structure

2 Vehicle Access and Tyre Pressure Monitoring Systems

The aim of Chapter 2 is to introduce the reader to the basics of the Vehicle Access (VA) and Tyre Pressure Monitoring topics. An overview of the possible system types and their functionality is presented with special care and reference to RF (Radio Frequency) theme, covering in this regard the relevant system parts.

Vehicle Access Systems and Tyre Pressure Monitoring Systems have got two main common advantages. They build not only convenience related functions, but contribute to the travelers and traffic safety.

From technical point of view, VAS as RKE or PASE and TPMS may be separate, fully autonomous. In such case all the parts are individually designed for each system. This means that all systems work independently from each other and have components dedicated solely for one application use. However, if a vehicle comprises more than one of the mentioned systems, such solution is not optimal. For a number of reasons it is favourable to combine these systems in order to share electronics and utilize some electronic parts as common. Among the reasons are the reduction of vehicle network complexity, cable routing, simplification of assembly processes, reduction of space demanded for electronics implementation and therewith weight reduction and resulting from all of these cost optimization. For all the noted points vehicle manufacturers are incorporating more and more functions into integrated parts and linked systems. Regarding TPMS, RKE and PASE, the most favourable solution is to merge the systems - the cross link is to be realized via shared RF-receiver. The functionality of each of the systems as well as their composition is described in the following subchapters.

2.1 Remote Keyless Entry

RKE is an abbreviation for Remote Keyless Entry. Remote Entry means that the system guarantees remote access to the automobile. The word "Keyless" is actually a misnomer since an electronic key fob duplicates all the features of a standard mechanical car key with the added convenience of operating the power door locks and further functions from a certain distance. A small traditional mechanical key is usually still included for unexpected cases (e.g. when the key fob is out of charge because of a flat battery) see Fig. 2.1. The need to physically manipulate a key into a door lock is eliminated.

An electrical key fob is equipped with several buttons dedicated to locking and unlocking the vehicle. Additional functions such as opening the trunk, closing the windows and sunroof etc. may also be included. Some remote keyless fobs also offer an additional panic button which activates the car alarm as a standard feature. Whenever a driver pushes one of the buttons and therewith induces a signal, the data is encoded and sent by the transmitter to the receiver positioned in the vehicle. In the next step data is decoded in order to determine what the demanded function is. Most of the systems imply only simplex (one way) communication and when the vehicle releases the door locks the operation is over. In other cases when duplex (bi-directional) communication is foreseen, the vehicle transmitter sends an acknowledgement signal back to the key fob. In such a case confirmation possibility of performed function is given - usually either blinking LEDs or a small key display serve as an interface for the user as in [23].

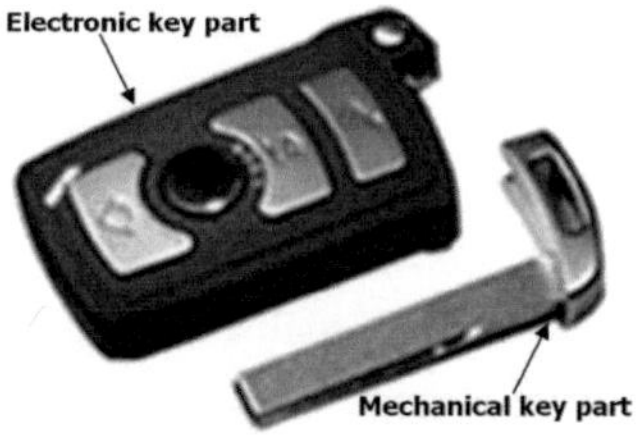

Figure 2.1: Electronic key fob

The installation rate for RKE systems in new vehicles is more than 80% in North America and more than 70% in Europe [24], [25], [1].

2.2 Passive Start and Entry System

PASE stands for PAssive Start and Entry and describes an innovative vehicle access and start system. The word "Passive" refers to the fact that a user does not have to act in any way in order to unlock the car nor to start the engine. This is contrary to the common access systems as for example RKE and therefore this is the most significant difference from the user point of view.

PASE allows automatic locking, unlocking the vehicle doors or trunk without pressing any key button nor even taking the key out of a bag or pocket. The engine is started by just pushing the ignition button. The car owner has only to carry the key, which has got the function of identifying the right user. Therefore it is very often called an Identification Device (ID), Identification Card (IC) or Access Card (AC). As the user approaches the vehicle and is within less than few metres distance, the vehicle automatically recognizes the identification device. In the next step the driver has to pull simply the door handle to access the vehicle. Only when IC is detected to be in the vehicle interior, the engine may be started by pushing an ignition button. ID card is authenticated; immobilizer and steering column lock are automatically released so that the engine can start. The vehicle is locked when the owner walks away.

Communication between the vehicle and the ID is realised at 125 kHz, leading to a high penetration and virtually no reflections, due to the large wavelength $\lambda > 2000$ m. As the communication between the LF-transmitter (Low Frequency) in the car and the identification provider (card owner) can only take place within the coverage range without any particular actions of the car owner, it is very important that the coverage range around the vehicle is well limited and controlled. The car access should be ensured only if the owner is within a few metres distance, typically about 1.5 m around the car.

2.3 Basic Approaches to Tyre Pressure Monitoring

TPMS stands for Tyre Pressure Monitoring System. As to derive from the system name, it is an electronic system designed to monitor the apparent air pressure inside all pneumatic tyres on a motor vehicle. The system is sometimes simply called a run flat indicator as it indicates if any tyre is lacking pressure. Argument for this system is the fact that pressure drops even to 30% are not remarkable for human eye. The difference between properly inflated (Fig. 2.2b) and underinflated tyre (Fig. 2.2a) is very difficult to state. Comparing however the cross sections of under inflated (Fig. 2.2c) and properly inflated tyre (Fig. 2.2d) it becomes very remarkable that the adhesion is strongly decreased.

(a) Underinflated tyre (b) Properly inflated tyre (c) Cross section of underinflated tyre, road contact (d) Cross section of properly inflated tyre, road contact

Figure 2.2: Difference between properly inflated and underinflated tyre

For this reason drivers are very often not aware of possible danger and hazard they are exposed to, which is associated with tyres. These are designed for use at their specific recommend pressure. In case of under inflation, the tyre is exposed to increased deformation in the wall region as it concentrates the load upon the tread shoulders. In the end effect tyre looses proper contact with the ground, and therewith the vehicle can get out of the drivers control. Further disadvantages and dangers are: increased tyre wear, higher fuel consumption, longer braking distance and also higher probability of a sudden tyre puncture.

Knowing all the dangers linked with under inflation, it is very important to stress that pressure loss is a natural and continuous process. On average, a car tyre loses about 0.1 bar air pressure per month [26]. Most of the motorists are not aware of this fact. They never or very seldom check the tyre pressure with one exception this is when changing winter to summer tyres. Therewith an automatic pressure measurement system is of very high importance for individual and traffic safety.

TPMS offers the following features:

- Tyre pressure measurement, ablility to detect under-inflation of less than 25 %.
- Continuous pressure monitoring both during vehicle movement and at the standstill.
- Early warning (< 5 s) for the driver, in case of pressure loss or tyre failure.
- Prevention of error while setting nominal pressure value.
- Automatic identification of location of wheels (left / right -font and rear).

From functional point of view, TPMS is divided into indirect and direct systems [27], [28]. Fig. 2.3 [29] shows the main components and data transmission paths of both systems.

2.3.1 Indirect Tyre Pressure Monitoring System

Indirect TMPS (ITPMS) uses information from other in vehicle integrated systems in order to determine the actual tyre pressure. Physical pressure sensor is inexistent in this type of TPMS. The information on pressure in all tyres is gained through the analysis of the wheel speed. Hardware parts such as wheel speed sensors as well as control unit are shared with ABS (Anti-lock Braking System). The only dedicated part of the system is the appropriate software.

The operation method of an indirect TPMS is based on the fact that under-inflated tyres have a slightly smaller diameter than a correctly inflated tyre. The difference between properly and insufficiently inflated tyres is optically very difficult to state. Nevertheless, few physical parameters may be easily estimated as for example the angular wheel velocity. In order to overcome the same distance an under-inflated tyre has to rotate at a higher angular velocity.

Main advantage of this system type is the moderate cost due to no need of applying any additional hardware except from user interface. Disadvantages, however, such as low sensitivity of the system and no information which wheel drops pressure unless intricate wheel localization, make this system to be of minor interest. As there is no need of radio based communication between system parts, it is not in the focus of this work.

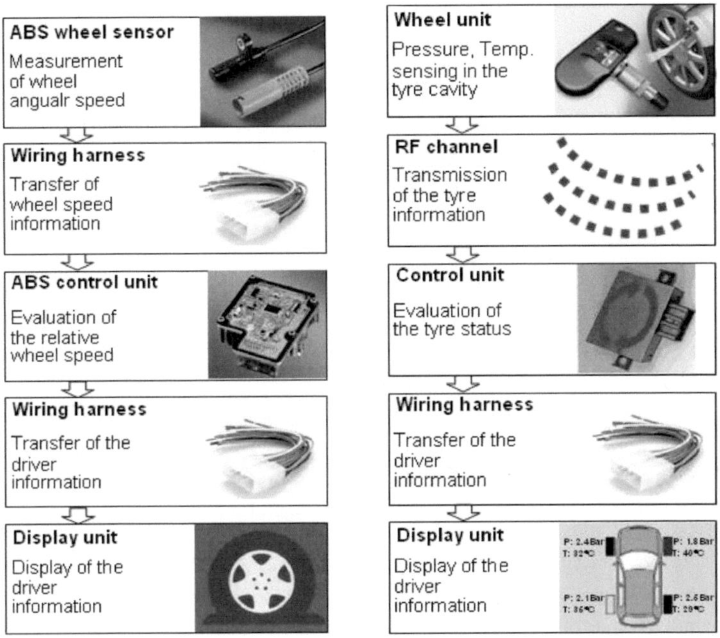

(a) Indirect Tyre Pressure Monitoring System (b) Direct Tyre Pressure Monitoring System

Figure 2.3: Comparison of Indirect and Direct Tyre Pressure Monitoring Systems

2.3.2 Direct Tyre Pressure Monitoring System

The name for the Direct TPMS derives from the fact that the pressure sensor is integrated directly in the wheels of the vehicle. This sensor, often named Wheel Unit (WU), is solely devoted and specially designed for TPMS with the functions of direct pressure and temperature measurement, contrary to ITPMS. The functionality of the system is depicted in Fig. 2.3b. The key component is the Wheel Unit located in the tyre cavity. The major design challenge is bound with this part. Detailed description of the WU will follow in subsequent chapters.

Radio Frequency transmission is of most importance for the overall performance of the whole system. The data is transmitted via wireless channel to the Control Unit (CU). Due to positioning within the vehicle body the CU is less demanding in its design as the WU. There are two options for the RF reception antenna. It is either integrated in the CU body or it is an external antenna and a connection to the CU is provided. Because of the cost of the parts and also the cost of assembly process, the internal antenna is the preferred solution. The CU processes the received information.

Finally, the warning messages and optionally the quantities measured in the tyres are forwarded via wiring harness to the display unit. Depending on the vehicle OEM, a variety of devices and user interface strategies may be employed. From a simple acoustic or light signal to very advanced combination of a display.

2.4 Functional Architecture and Frequency Allocation

The functional architecture describes main system components along with their mutual interaction. From structural point of view it was already mentioned that it is advantageous to merge VAS and TPMS. Resulting core and peripheral system components are described underneath. Joined VAS and TPMS system diagram is presented in the Fig. 2.4.

The most important logical blocks of each interacting system component are depicted. Central - core part in which the functionality of all systems is merged is the RF-receiver with Control Unit (CU). It comprehends electronic hardware as well as software functionality blocks for all three systems. Connectivity to distant, peripheral units specific for each of the systems is assured with the help of RF and LF Interfaces with equivalent antennas, receivers / transmitters and microcontrollers. Systems communicate both in LF and RF range. LF communication takes place at the frequency of 125 kHz, whereas RF transmission takes place at higher frequencies in ISM (Industrial Scientific Medical) bands.

Peripheral units are respectively for VAS either PASE Identification Device or RKE key, and for TPMS the Wheel Unit. In this thesis described system is determined for operation within European countries and so between CU and VAS peripheral components the frequency of 868 MHz is used. Data transfer from TPMS WU is carried out at 433.92 MHz. Further frequencies dependent on the implementation region are respectively 315 MHz, 912 MHz for Japan and USA.

The CU receives WU information as well as RKE key information by means of the RF subsystem comprising antenna and RF receiver. In the second step, the received data frames are processed by microcontroller, the information is forwarded accordingly to further system blocks, either for TPMS or RKE. When the content of the information is approved, corresponding signals are send to the User Interface (UI). For TPMS last block is the UI where the information about tyre pressure is displayed and for RKE the power door locks.

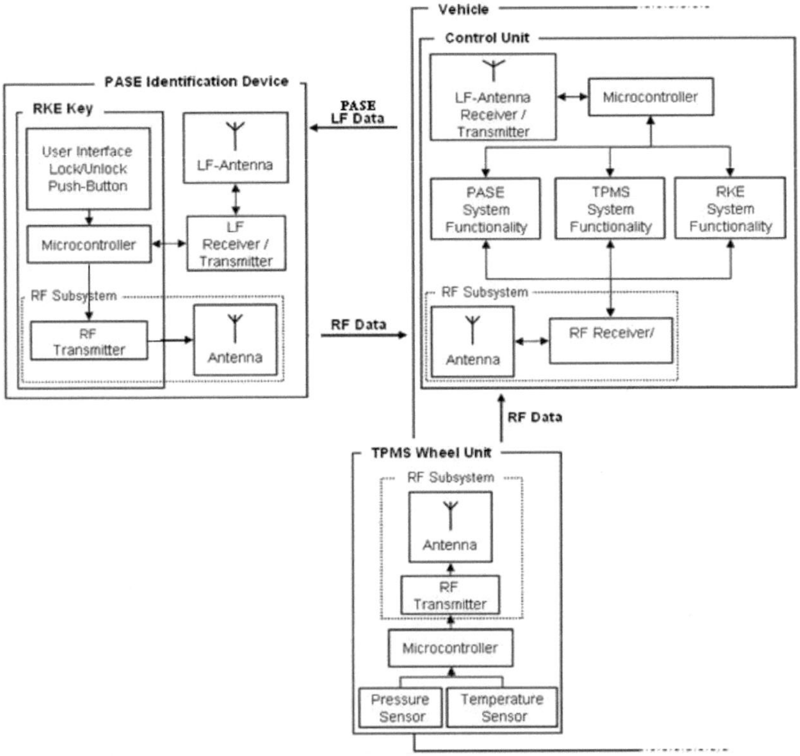

Figure 2.4: Integration of Tyre Pressure Monitoring System, PAssive Start and Entry and Remote Keyless Entry

Both in RKE and TPMS the information signal originates from external transmitters. These are in the first case the hand-held key fob and in the second the Wheel Unit inside the tyre cavity. From RF point of view both devices have similar hardware architecture. The only difference is that the transmitted signal is a carrier for another kind of information and it is initiated in a different way. Holding a key fob person approaching the vehicle is the decision taker whether and when to command the door opening. To emit the order signal one of the buttons on a key fob have to be pushed. Depending on a number of functions the key fob has got one or more push-buttons and optionally a display. On the contrary the

WU is initialized via the motion sensor, from now on it transmits the information in regular time intervals.

Some of the vehicles may be equipped with a more advanced and comfortable vehicle entry system PASE (PAssive Start and Entry), where driver may enter the vehicle without any explicit action. Authorization is granted by carrying an appropriate Identification Device (ID). In this case RKE key fob is extended with additional operational blocks such as LF transmitter and related antenna. LF communication that occurs between the vehicle and ID is unidirectional.

Described above system functional architecture entails advantages for further hardware parts, which have to build in the vehicle only once. These are housing, connector, power supply, car bus, micro-controller, LF drivers and RF antenna. All the devices that are relevant from RF point of view will be described further on.

2.5 RF Relevant Parts of the Systems

Relevance describes to what extent objects are important for a solution of a defined problem. With RF relevant parts such vehicle and system parts are meant which influence functioning of the antennas to be installed and modify wave propagation. Moreover, under RF relevant parts of the system not only the integral hardware parts of RKE, TPMS, and PASE are to be understood but also the parts that have influence on the system performance dependent on the mounting position of the transmitter and receiver.

Summing up and following this logic, RF relevant parts of VAS and TPMS may be divided into two groups: firstly system integral parts, secondarily further elements, which do not belong directly to the described systems but are vehicle or environment specific. Both metallic and dielectric elements have to be taken into account.

Integral, RF relevant merged system parts are:

- RF receiver with adequate antenna.

  The antenna may be integrated into the receiver module as in [2], or it may be for example an external antenna like a film antenna integrated into the windshield as in [30]. There might be one or more receiver positioned within the vehicle. The developed antenna will be in depth discussed in Chapter 4 dealing with antenna development.

- Tyre pressure sensor unit already before described as Wheel Unit (WU).

  The positioning of the Wheel Units is determined by their functionality. It has to be within the wheel cavity. Additionally it precisely defines further RF relevant but not system integral parts.

- RKE key and / or PASE ID

  RKE key or PASE ID is a hand-held device. During the development process the influence of human hand is always analysed. The design of the key module falls beyond the intent of this thesis. A functional RF key module is assumed.

Indirect, RF relevant system parts that influence system functioning are:

- Vehicle body and equipment
- Rim
- Tyre (materials, construction, motion)
- Ground (ground type and condition)

It is important to mention the reason why all listed elements affect the functionality of the parts that directly belong to and build VAS and TPMS systems and in which way this occurs. Electromagnetic interactions between integral and indirect system parts have to be understood. following issues are necessary for the optimization process of RF properties:

- Simulations of the influence of tyre components on the wave propagation help choosing adequate antenna structures for best performances of the RF systems.

- Rims and tyres influence the antenna integrated within the WU, in such a way that the electrical antenna parameters such as impedance and radiation pattern are modified.

- The vehicle body together with the car equipment considerably affect the wave propagation channel from the tyre pressure sensor to the receiver depending on the location of the receiver inside the vehicle [9], [15].

- The ground as asphalt, concrete and gravel likewise affect the wave propagation.

Consequently, the hardware parts as well as the next environment of the hardware components are important for a correct RF analysis. All these elements were taken into account whilst design process. Based on the simulation and measurements a high system reliability and robustness may be achieved with high accuracy of RF behaviour prediction. Further Chapters will give a deep insight in this problematic, especially concerning the simulation process and element mutual influence.

3 Passenger Vehicle Tyre Classification

Goal of Chapter 3 is to introduce an approach how to classify tyres. In order to find out which mechanical and material properties influence tyre RF properties, knowledge about different tyre types and their structure is needed. Chapter 3 together with Appendix A deals with the tyre classification from mechanical and RF point of view. It has to be defined which parts or layers of the tyres have to be examined. Therefore, firstly tyre types and geometry are introduced. Following assignment based on market overview was made with emphasis on most popular tyres with highest sale quotas. All information, gained in this manner, forms the base for the first simplified RF simulative tyre models described in detail in Chapter 6.

Tyre (AE *tire*) is an elastic shock absorbing, ring-shaped covering that is assembled around the wheel surface. Tyres feature very complex structure. They have to withstand very harsh environmental conditions, rapid temperature changes, high loads, stress, mechanical shock and deformation. Moreover exposed to these terms tyres have to show long life cycle. First tyres, however, were bands of iron or steel placed on wooden wheels, used mostly on carts. Pneumatic tyres, as they are known nowadays, were introduced relatively lately, in the 19th century. These consist of a flexible elastomer material such as rubber reinforced by fibre and wires. Growth of tyre market is strictly bound to the automobile industry and so the first companies manufacturing tyres in big scale aroused in the early 20th century. Demand on tyres and their development increased together with vehicle success and popularity. Today over one billion tyres are produced annually [31].

The variety of tyres present on the worldwide market in huge and therefore also the differences between the tyres. There are however only less than ten leading manufacturers commanding most of the global market share. Big suppliers deliver very high quality products, support vehicle OEMs, whereas the smaller ones address mostly low price aftermarket segment. In order to be competitive and increase safety, tyre manufacturers vie for preparing new tyre designs. There are multiple parameters upon which tyres are described. Important constraints are rolling resistance, durability, ride comfort, steering precision, stability, weight, life expectancy and behaviour on wet and icy roads. It is important to notice that all these parameters are related to mechanical tyre performance and properties.

Dependent on the deployment rubber compounds with different chemical additives are used in the manufacturing process. Generally for winter tyres more elastic rubber than for summer tyres is used, however exact material components are unknown. This information is of highest secrecy and confidentiality for each tyre manufacturer.

Summing up tyres are very well characterized from mechanical performance point of view, whereas material properties, especially electric properties such as permittivity and loss factor were until now to the authors knowledge unknown.

3.1 Overview of Tyre Types

Before discussing tyre structure in detail, different tyre types have to be shortly presented. There are three major construction types: (1) Slick tyres, (2) Conventional tyres, (3) Self Supporting Runflat (SSR) tyres.

Slick tyres are very well known to ones who have interest in racing sports. The main optical difference to conventional tyres is the lack of profiled tread. Slick is another word for smooth, slick tyres have very smooth tread (Fig. 3.1a).

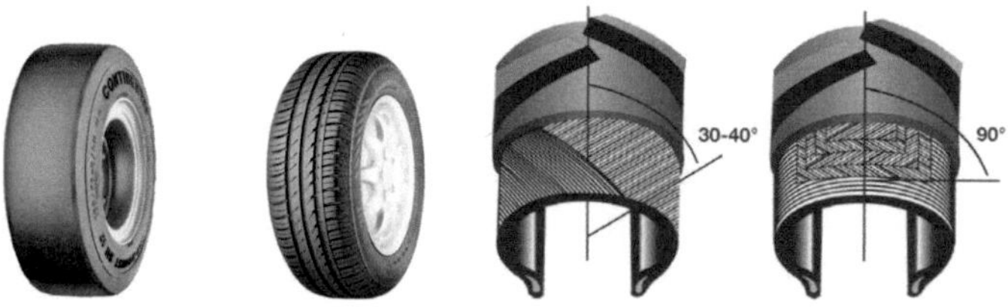

(a) (b) (c) (d)

Figure 3.1: Comparison of tyre type and comparison of reinforcement construction (a) - Example of slick tyre, (b) - Example of conventional tyre, (c) - Diagonal tyre construction, (d) - Radial tyre construction

The aim is to provide the largest possible contact patch to the road and maximize traction for any given tyre dimension and very fast speeds. The drawback is, that this kind of tyres are not suitable for usage on common road vehicles. In case of wet, icy or snowy road these tyres loose good ground contact.

Contrary, conventional tyres (Fig. 3.1b) must be able to operate in all weather conditions and usually requirement for extreme velocities over 200 km / h is not demanded. In order to drain off water, snow and sand tread is grooved. This is a very important feature and therefore manufacturers examine different profile geometries. Further difference is an additional reinforcement. Tyres are strengthened by fibre web. Fibres may be arranged in two ways: diagonally and radially with respect to tyre centre line. In both cases this reinforcement is placed over the whole tyre structure from the inside to the outside bead. Diagonal tyres have net (Fig. 3.1c) where fibres are directed diagonally at angle of 30 ° to

40 ° to tyre centre line. There are two layers of in this way formed carcass. The superimposed layers run in opposite direction to form a cross pattern. In radial tyres (Fig. 3.1d) the carcass layers run at angle of 90 ° to tyre centre line. Additionally directly over the carcass under the tyre tread there is a multilayered cord or steel belt. Wires in the belt are oriented between 15 ° and 25 ° to the tyre centre line. Most of the nowadays implemented tyres are radial tyres.

Another tyre type that has grown in popularity are Self Supporting Runflat tyres, often called in short Runflats or Zero Pressure tyres. SSR tyres have got, in comparison to conventional tyres, an additional side wall strengthening (Fig. 3.2a). When tyre is punctured - flat, these side walls support the whole tyre structure (Fig. 3.2b). SSRs are designed to resist the effects of deflation and to enable the vehicle to continue to be driven at reduced speeds for limited distance [33], [34].

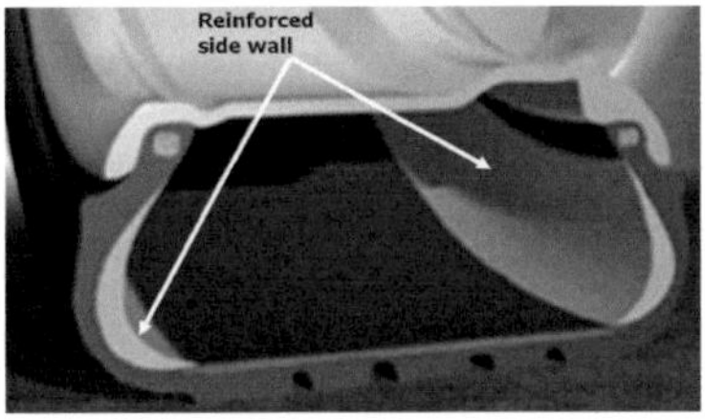

(a) SSR tyre with reinforced sidewall

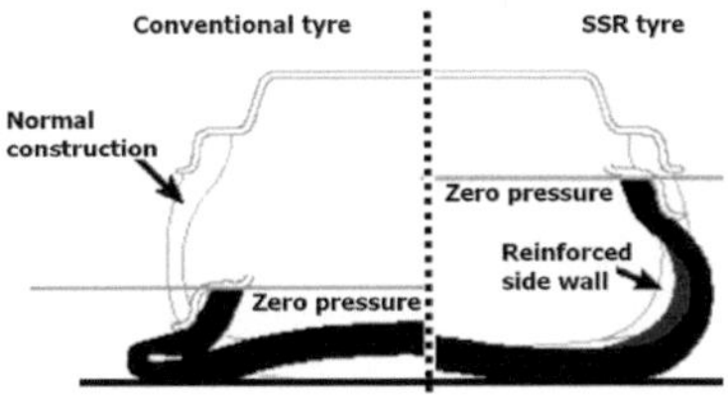

(b) Comparison of conventional and SSR tyre in case of puncture

Figure 3.2: Distinction between conventional and SSR tyre

Summing up, there are two basic types of conventional tyres, however due to advancement in vehicle technology and higher reachable speeds, tyres with radial construction dominate the market. The slick tyres in comparison to conventional tyres are of relatively simple construction. It is important to mention that SSR tyres do not have yet a very big market share but in all vehicle were they are implemented a TPMS is demanded.

3.2 Geometrical Tyre Structure - Main Components

Years of development and research brought tyres to be products with a very high complexity level. Standardized tyre classification criteria upon markings on tyre side, are described in Appendix A.1. Chapter 3.2 gives insight to the tyre build up and components, which being part of the complex structure influence the electromagnetic tyre properties.

Main compound - natural and synthetic rubber is mixed with such additives as sulphur, carbon black and other solvents to form base for tyre structure [36]. Following the most modern tyre development progression, up to 25 different tyre parts and up to 13 different rubber mixtures may be used [37]. Never though, tyre components are usually set in a predefined way, they are different (contain diverse ingredients in differing amounts) for each manufacturer.

Tyres are composite products. The tyre consists of up to 9 different layers - depicted in Fig. 3.3. Each of the depicted layers is of high importance for an overall mechanical but also RF tyre performance. It is shown how much each of these components influences the WU radiation characteristic.

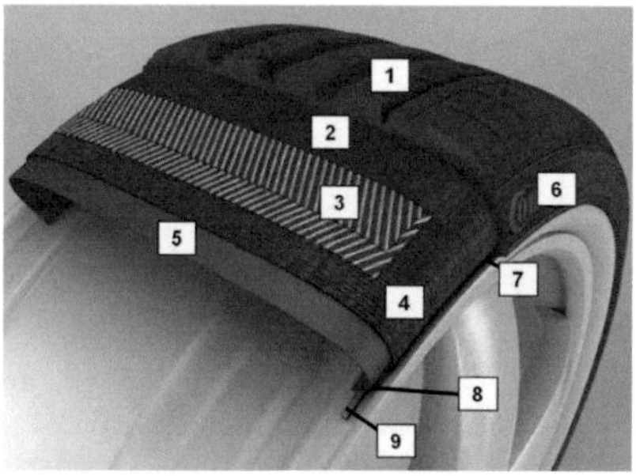

(a) Tyre structure 3D

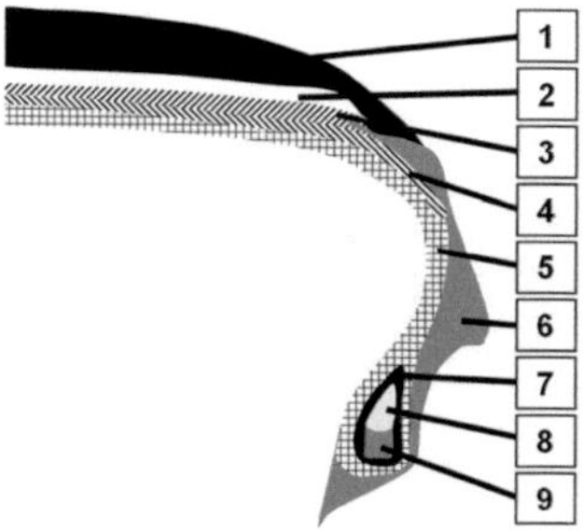

(b) Tyre cross section

Figure 3.3: Tyre structure: (1) tread, (2) wire belt, (3) carcass first layer, (4) carcass second layer, (5) inner liner, (6) side wall, (7) bead reinforcement, (8) apex, (9) bead core

Besides the chosen materials, specific vulcanization procedure and tyre construction, the capabilities and performance of each tyre depend strongly on the inflation pressure. For this reason TPMS grows in importance. Representative tyres are chosen for the RF investigations, therefore a market overview in necessary.

3.3 European Market Players / European Tyre Manufacturers

The European passenger car tyre market represents 33% [38] of the world total market. Manufacturers with biggest market shares are: Bridgestone, Continental, Dunlop, Goodyear, Michelin and Pirelli. These are taken into account for the analysis within the scope of this work.

In order to realize a merged classification for tyres of different brands, knowledge has to be gained upon which characteristics this classification should be realized. Therefore firstly general tyre description will be expound and consequently tyres classified upon mechanical and geometrical parameters.

3.4 Tyre Classification and Overview

For tyre classification conventional and SSR tyres are taken into account. An overview of summer and winter tyres existent on the market (in year 2007) was made (see Appendix A.3).

Tyres of the manufactures with the biggest market shares described in Chapter 3.3 were taken into account. All manufactured tyres were compiled within tables from Fig. A.2, Fig. A.3, Fig. A.4 and Fig. A.5. Resulting in 1732 conventional summer tyres, 817 conventional winter tyres, 202 SSR summer tyres and 91 SSR winter tyres. Products of each manufacturer were coded with different pattern / colour. In all graphics tyres are arranged upon speed index (horizontal table direction, rising order) and size (vertical table direction, rising order).

The structure of tyre market is very well depicted with the realized and presented classification. It is noticeable that smaller tyres have lower speed index, whereas larger tyres feature higher speed indexes. Moreover with the help of this study it is easy to conclude which tyres are most popular, of which size and which speed index. A brief extract is presented in Fig. 3.4 and Fig. 3.5.

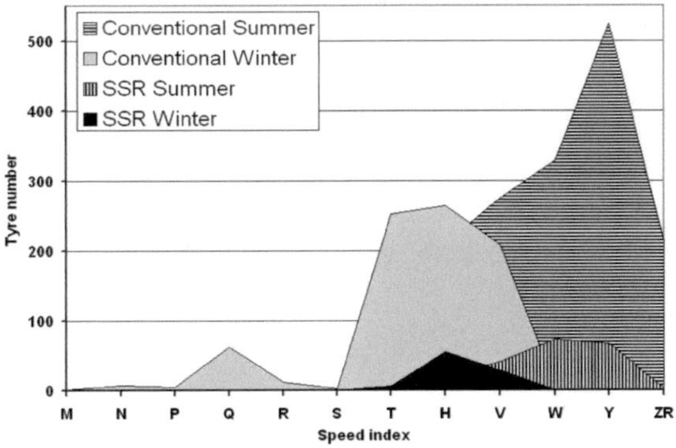

Figure 3.4: Market share of conventional and SSR tyres in respect to speed index

The above shows how many tyres of different rim diameters are offered on the market. Most popular conventional tyres have the diameter of 15 to 18 inch. In segment of SSR tyres diameters smaller than 16 inch are not available.

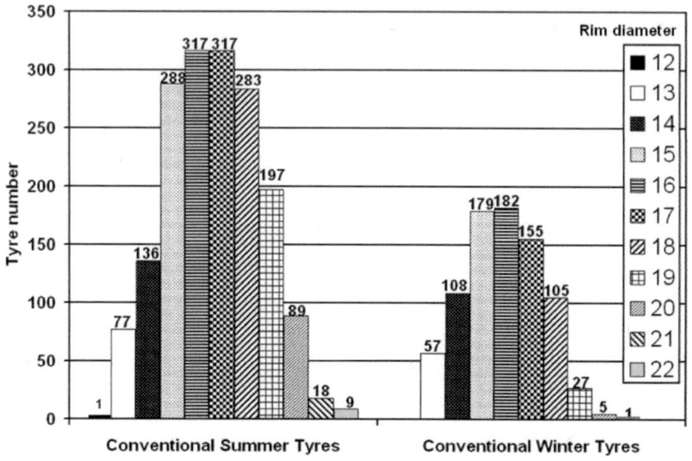

(a) Conventional tyres

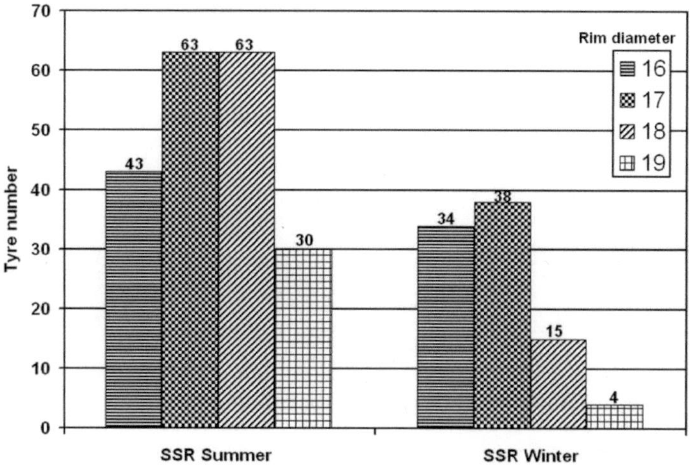

(b) SSR tyres

Figure 3.5: Market share of summer and winter tyres in respect to rim diameter

Additionally comparing tyre types it becomes very transparent that most tyres offered on the market are conventional summer tyres followed by conventional winter tyres. Market share of SSR tyres is still rather small. Interpreting the dependency of speed index it is remarkable that summer tyres, independent whether conventional or SSR, are allowed to run faster than all winter tyre types. In order to hold further investigation representative (not to investigate rare parts), the decision was taken to analyse most popular tyre segment with the following size:

- 195/65 R15 (further called small tyres)
- 245/40 R18 (further called big tyres)

Both summer and winter tyres in above listed sizes were chosen to be analysed, as well as a selection of SSR tyres. Full table with exact tyre names, brands is available in Appendix A, Table A.4.

Though, before constructing appropriate tyre model for simulation purposes, below mentioned goals have to be fulfilled in order to gain necessary knowledge about differences between tyres from RF point of view. Chosen tyre selection is optimal in order to follow proximate goals in further paragraphs. Understanding differences and similarities between:

- Tyres from diverse manufacturers.
- Summer and winter tyres.
- Tyres of various speed indexes.
- Conventional and SSR tyres.
- Different tyre sizes.

Moreover it is important to investigate the tyre geometry having on target the simplification of its electromagnetic model. This may be undertaken after gaining the knowledge about RF relevance of each tyre component from Fig. 3.3.

3.5 RF Tyre Classification upon Radiation Pattern

The radiation pattern is a graphical representation of the antenna radiation properties as a function of space. The depicted value is the relative field strength transmitted from or received by the antenna. It refers to the directional (angular) dependence of radiation from an antenna or other source [39], [40], [41].

For complete, full description of antenna radiation properties a 3D representation is demanded. Unfortunately 3D measurements have some drawbacks as they are very time consuming, require special equipment and big anechoic chambers. Therefore it is common to describe the radiation pattern with three principal plane patterns. These are obtained by making three cuts through the 3D pattern.

Typically radiation pattern is characterised in one horizontal and two vertical planes. Accordingly *xy* ($\theta = 90°$), *xz* ($\varphi = 0°$) and *yz* ($\varphi = 90°$) plane. Corresponding rectangular and spherical coordinate system is presented in Appendix B (Fig. B.1a).

For description of antenna performance in form of radiation pattern C many different quantities may be used. These are for example electrical field strength E, radiated power P_{rad}, efficiency η and directivity D. Another useful measure describing antenna performance is the antenna gain G. It takes into account both the efficiency as well as directional capabilities of radiating structure (Eq. 3.1).

$$G = \eta D \tag{3.1}$$

Contrary, directivity D (Eq. 3.2) is defined as a measure that takes into account only the directional radiation properties of the antenna and therefore it is only influenced by the antenna pattern C.

$$D = \frac{4\pi}{\int\limits_{\varphi=0}^{2\pi} \int\limits_{\theta=0}^{\pi} C^2(\theta, \varphi)\ sin\theta\ d\theta\ d\varphi} \tag{3.2}$$

Radiation pattern C (Eq. 3.3) is defined with ratio of electrical ($\vec{E}$) or magnetical ($\vec{H}$) field strength on sphere surface and there occurring maxima ($\vec{E}_{max}$, $\vec{H}_{max}$):

$$C(\theta, \varphi) = \left. \frac{\left|\vec{E}(\theta, \varphi)\right|}{\left|\vec{E}(\theta, \varphi)\right|_{max}} \right|_{\substack{r=const \\ r\to\infty}} = \left. \frac{\left|\vec{H}(\theta, \varphi)\right|}{\left|\vec{H}(\theta, \varphi)\right|_{max}} \right|_{\substack{r=const \\ r\to\infty}} \tag{3.3}$$

Efficiency (Eq. 3.4) expresses the relationship between radiated P_{rad} and input power P_{in}. It is a purely inherent property of the antenna [39], impedance and polarization mismatches do not contribute to efficiency. However, as antenna specific losses cause the fact that radiated power is always lower than input power, efficiency may be seen as attenuation factor for the input signal.

$$\eta = P_{rad}/P_{in} \tag{3.4}$$

There are two basic methods to perform efficiency measurement:

- Wheeler cap [42].
- 3D far field radiation pattern measurement.

The first of the above mentioned methods is suitable for small antenna modules, the second, as already mentioned before, is very time consuming.

Therefore in order to characterise the radiation pattern properties from different tyres, gain measurement was conducted with Two-Antenna-Method [39]. Block diagram of the measurement setup is depicted in Fig. B.4. Measurement of 2D radiation pattern was the most optimal as well the most accurate way to determine whether there are any differences between tyres from RF point of view.

Unknown gain of Device Under Test G_t, here tyre, is calculated according to Friis Equation [43]. It is expressed with following Formula 3.5, where λ is the wavelength, d is the distance between DUT and test antenna, respectively G_r and P_r are the gain and received power by the test antenna.

$$G_t = \frac{P_r}{P_{in} G_r} \left(\frac{4\pi d}{\lambda} \right)^2 \tag{3.5}$$

Friis equation describes the propagation behaviour of a signal transmitted by one and received by second antenna. This Link Budget may be expressed in dB with Eq. 3.6.

$$G_t = P_r - P_{in} - G_r + 10log\left(4\pi d/\lambda\right)^2 \tag{3.6}$$

All tyres were oriented during the measurement in the same way. Positioning of the tyre in respect to the coordinate system is depicted in Fig. 3.6. Measured values are expressed in dBi related to the isotropic radiator (Fig. B.1b).

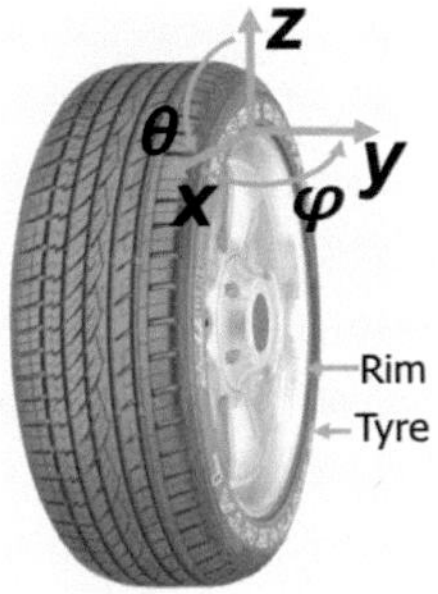

Figure 3.6: Tyre orientation for radiation pattern measurement

At this stage of the investigations, the WU, that actually originates the electromagnetic radiation, is to be treated as a black box. Tyre together with the WU is to be considered as one radiating entity under test. The influence of different WUs will be discussed in Chapter 5.

Classification criteria for the comparison of radiation pattern plots are the radiation pattern form, amplitude and eventually possible polarization dominance. For this purpose all graphs are depicted with the same scale.

3.5.1 Radiation Pattern Influenced by Conventional Tyres

The aim of the radiation pattern measurement influenced by conventional tyres is to see whether tyres of different brands influence the radiation characteristics in the same way. For this purpose WU with IFA antenna is positioned on the rim within each tyre cavity as in Fig. B.2.

Further questions for which an answer is required is whether there is any difference in the influence of winter and summer tyres as well as tyres of smaller and bigger size.

In order to answer the above queries, radiation pattern of chosen representative tyres with the same electromagnetic wave source within tyre cavity (test WU) was measured. Numerous characteristic curves and figures resulted from this investigation and therefore detailed results of this investigation are gathered in the Appendix B.2.

Chosen four tyre groups under test are:

- small (195/65 R15) summer tyres.
- small (195/65 R15) winter tyres.
- big (245/40 R18) summer tyres.
- big (245/40 R18) winter tyres.

It appeared that independent on the tyre manufacturer the radiation characteristic is influenced in a very similar way with similar radiation pattern characteristic. Therefore beneath only average values of the the achieved results are presented.

An average over all inspected summer and winter tyres was calculated. This in order to ease proving whether and to what extent winter and summer tyres differ.

Fig. 3.7 presents the comparison between small summer and winter tyres and accordingly Fig. 3.8 presents the comparison between big summer and winter tyres.

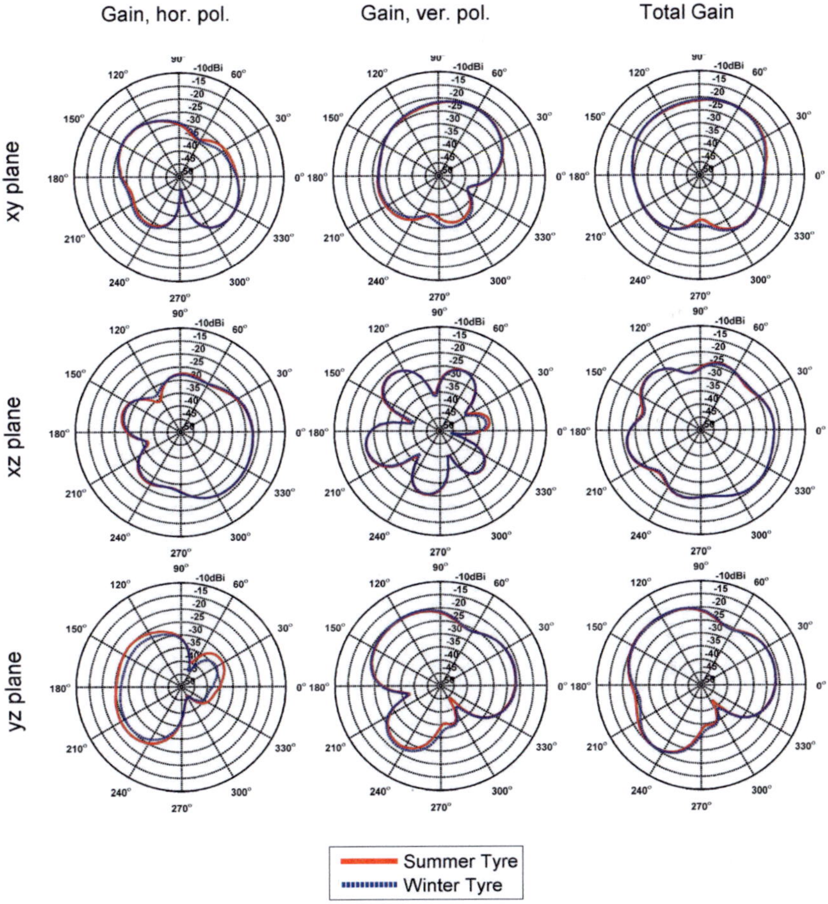

Figure 3.7: Comparison of average measured gain for conventional small (195/65 R15) summer and winter tyres

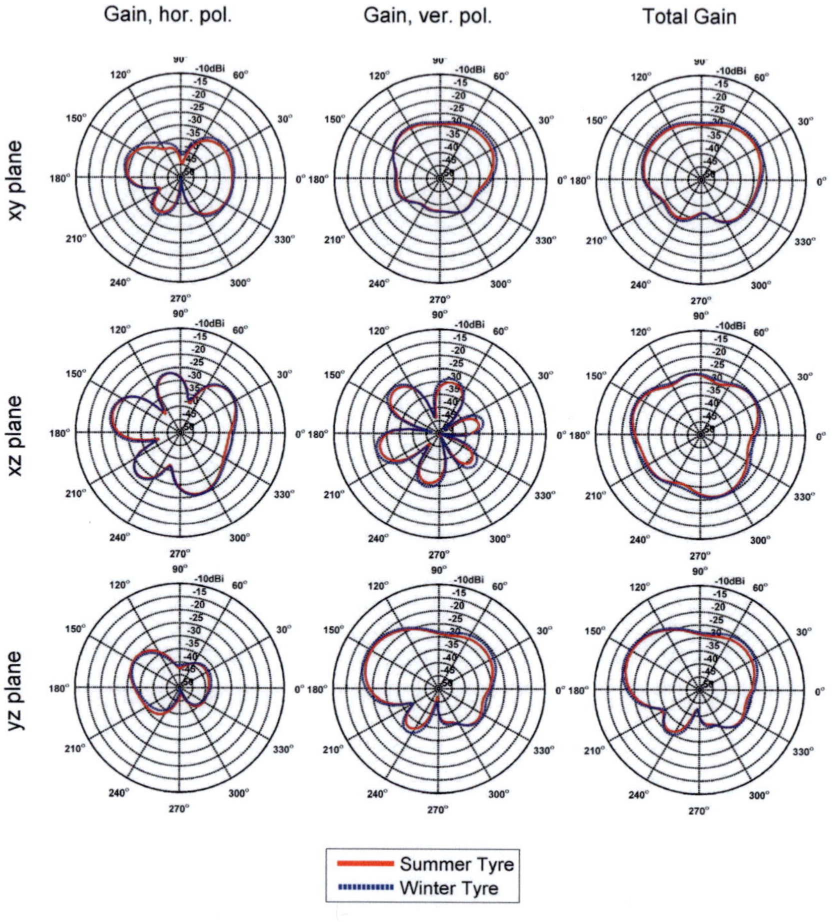

Figure 3.8: Comparison of average measured gain for conventional big (245/40 R18) summer and winter tyres

It is observed that there is small (of 1 dB to 2 dB) to no difference between the average values of the measured radiation pattern of small summer and winter tyres. The results for big summer and winter tyres show the same behaviour. In both cases it is clear that there is no difference between radiation pattern resulting from summer and winter tyres.

3.5.2 Radiation Pattern Influenced by Self Supporting Runflat Tyres

The comparison of all analysed SSR summer tyres is depicted in Fig. 3.9 and all SSR winter tyres in Fig. 3.10.

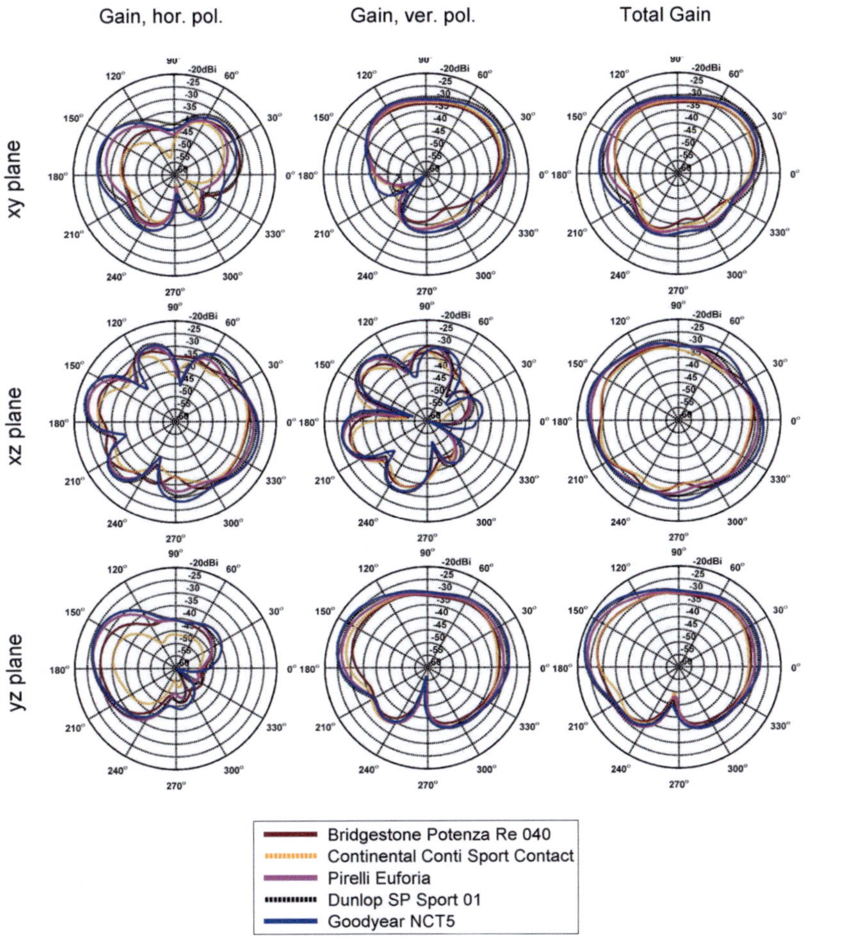

Figure 3.9: Measured radiation pattern of SSR summer tyres (245/40 R18)

Concerning summer tyres, the best tyre with highest gain values is the Dunlop SP Sport01. The worst tyre is the Continental Conti Sport Contact. The biggest difference between the average gain values is 3.4 dB (Table B.6).

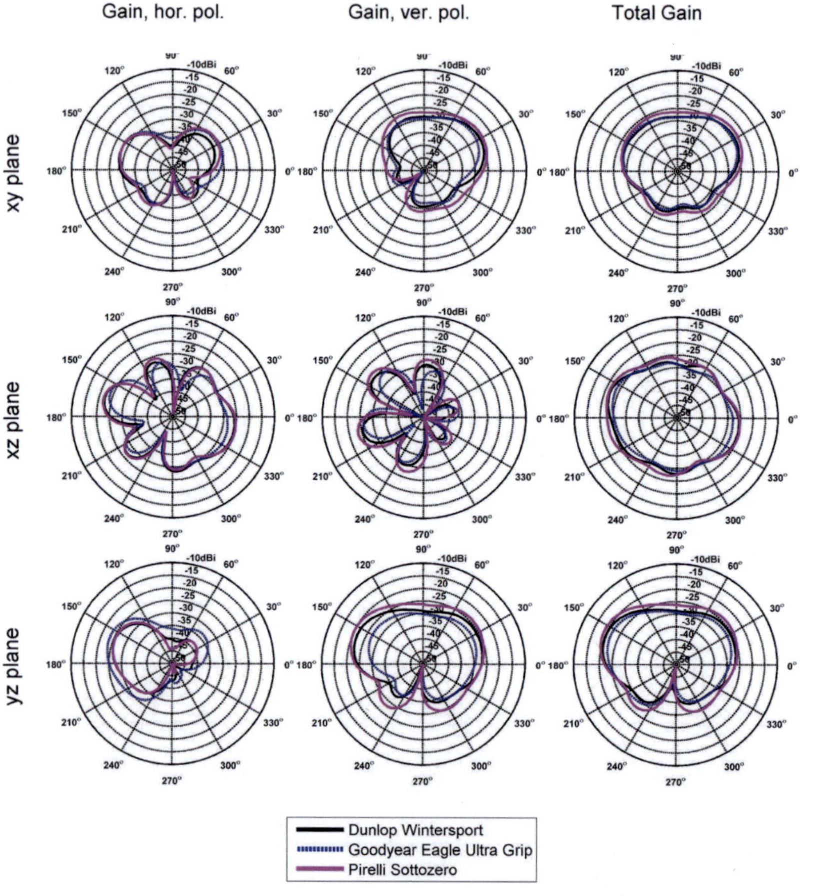

Figure 3.10: Measured radiation pattern of SSR winter tyres (245/40 R18)

Concerning winter tyres, the best tyre is Pirelli Sottozero, and worst is Goodyear Eagle Ultra Grip. The highest difference between the average gain values is 2.6 dB (Table B.5).

Summing up alike the case of conventional tyres, also all SSR tyres reveal the same radiation pattern form independent from the manufacturer and brand. Differences appear in the amplitude of the patterns.

3.5.3 Conclusions of RF Tyre Classification

Conducted RF tyre classification based on radiation pattern measurement, gain calculation and comparison of all measured tyres showed that:

- All measured conventional summer tyres show similar radiation pattern shape.
- All measured conventional winter tyres show similar radiation pattern shape.
- All measured SSR summer tyres show similar radiation pattern shape.
- All measured SSR winter tyres show similar radiation pattern shape.
- In each of above mentioned tyre classes, tyres of different manufacturers show discrepancy in radiation pattern amplitude. Therefore, it was possible to choose always the best and worst tyre candidate (taking into account radiation pattern amplitude).
- Radiation patterns of conventional summer and conventional winter tyres do not differ.
- Conventional and SSR tyres show different radiation pattern characteristics.

Above conclusions are meaningful both for characterization of tyres electrical properties as well as for tyre modelling. If measured radiation patterns of entities such as tyres with the same WU do not differ, this means, that also tyre electrical properties should not differ much. Consequently the same tyre simulation model may be used and only selected tyres will undergo material characterization process (Chapter 4).

In this sense conventional summer as well as conventional winter tyres will be represented in Chapter 6 with one common tyre model. SSR tyres due to another geometry (Fig. 3.2a) and therefore different radiation pattern require a separate tyre simulation model.

3.6 Influence of each Tyre Component on the Radiation Pattern

Having already the knowledge about tyre radiation characteristics and differences between different tyre types and brands, it would be favourable to analyse theoretically with means of simulation (further information in Chapter 6), which parts of tyre geometrical structure (Fig. 3.3) have most influence on radiation pattern and overall RF tyre performance.

For the investigation, variables of interest are permittivity ϵ_r and loss tangent $tan\delta$. In each step of the simulation process only one chosen parameter of interest was varied, whereas all further parameters were fixed. This resulted in numerous simulations. The advantage of such simulation procedure was that influence of each tyre layer and parameter was isolated.

At this stage of the investigations, when exact values of electrical parameters have not been measured yet, the range of values to be varied was chosen upon general knowledge in this field and some literature sources [44], [45]. Such choice guarantees plausibility of aimed investigation.

Electrical permittivity of air is close to that of vacuum, which is assigned to $\epsilon_r = 1$. For pure rubber materials the value is expected in the range of $6 < \epsilon_r < 10$. Tyres are however compound materials with different additional substances. One of the most popular additives is the carbon black with $\epsilon_r = 30$. Therefore, theoretical assumption for the investigations were values of $5 < \epsilon_r < 30$, simulation step of 5.

As rubber is a lossy dielectric material, apart from dielectric constant also the losses in the material have to be taken into account. These are expressed with help of loss tangent $tan\delta$. For the first estimations extreme values of this parameter were taken into account, ranging from $tan\delta = 0.001$ up to $tan\delta = 1$. The very low parameter value symbolises low loss materials.

Initial variables values are listed in underneath Table 3.1. Tyre under test is in size of 245/40 R18, naturally all simulations were conducted with the same electromagnetic wave source, this is the WU with IFA antenna. The influence of relative permittivity and of loss tangent is investigated.

Tyre Layer	ϵ_r	$tan\delta$
Tread	5	0.05
Inner liner	10	0.05
Sidewall	5	0.05
Bead reinforcement and Apex	12	0.05

Table 3.1: Initial values of ϵ_r and $tan\delta$

Fig. 3.11 depicts the influence of the permittivity values and Fig. 3.12 of the loss tangent for the tread rubber layer.

With the help of the above simulation it may be observed that the variation of ϵ_r of the tyre tread, does not cause any change in the radiation pattern. Differing loss tangent values $tan\delta$ in the tread, does not cause any change in the radiation pattern, neither the shape nor the amplitude change. This is due to the separation of this layer from the radiator (WU) with metallic wire belt (Fig. 3.3).

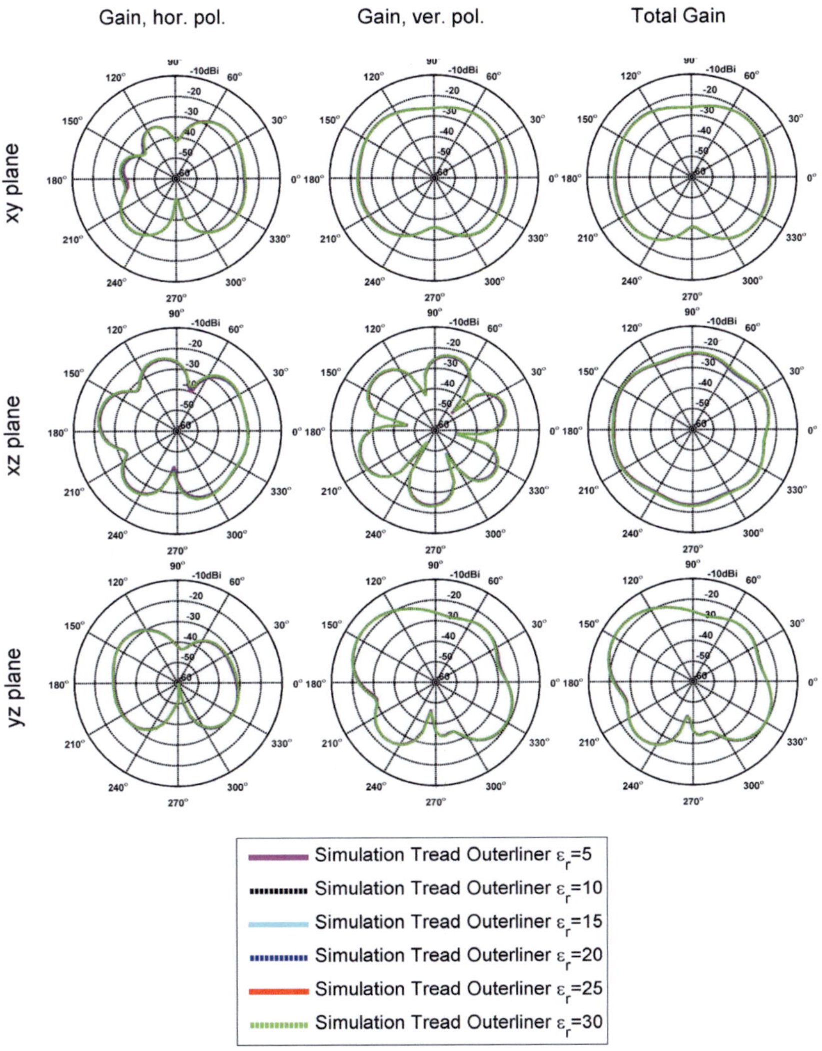

Figure 3.11: Radiation pattern simulation of 245/40 R18 tyre - variation of relative permittivity values of tread

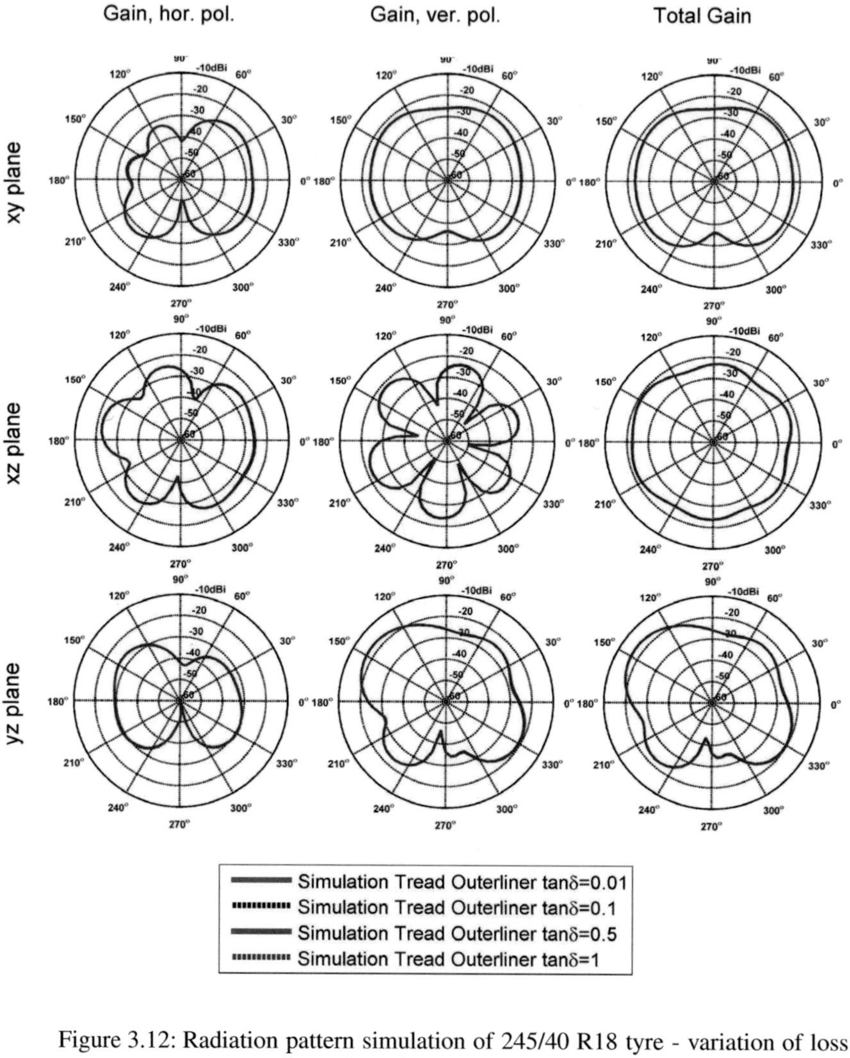

Figure 3.12: Radiation pattern simulation of 245/40 R18 tyre - variation of loss tangent values of tread

Fig. 3.13 shows the influence of the permittivity values for the inner liner.

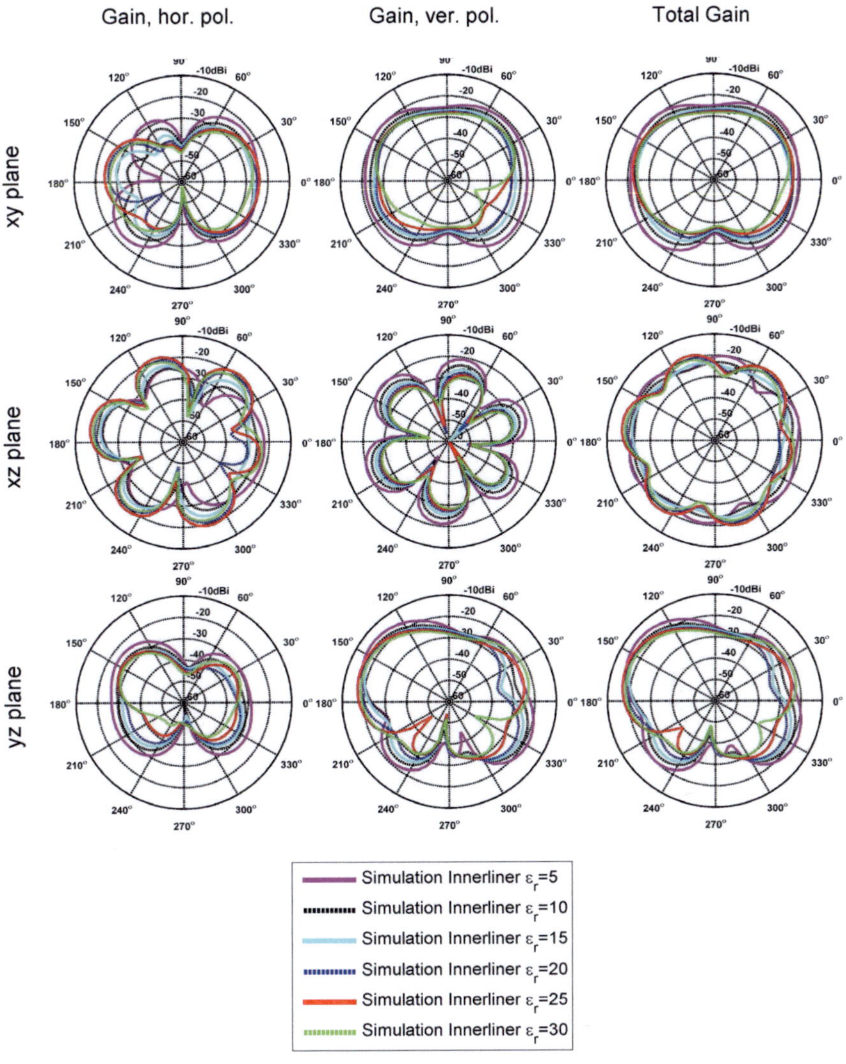

Figure 3.13: Radiation pattern simulation of 245/40 R18 tyre - variation of relative permittivity values of inner liner

Contrary to the tread of a tyre, the inner liner takes actively part in the radiation of the entity of WU and tyre. It is noticeable that the higher the values of relative permittivity are, the lower the amplitude of the radiation pattern. Change in the amplitude of the radiation pattern amounts max. 5 dB for lowest and highest simulated ϵ_r values. The radiation direction does not change significantly.

Fig. 3.14 shows the influence of the loss tangent values for the inner liner.

The variation of $tan\delta$ affects the radiation characteristics in similar way as for variation of ϵ_r. The higher the value of the simulated parameter, the lower the gain (Fig. 3.14). Though, maximum notable difference in the amplitude is of 3 dB.

Fig. 3.15 shows the influence of the permittivity values for the side walls.

The influence of tyre sidewall parameters is very remarkable. Variation of ϵ_r results in big changes of the radiation amplitude (changes up to 15.7 dB) as well as in radiation pattern shape. The lower the value of ϵ_r the stronger and more omnidirectional radiation characteristic.

Fig. 3.16 shows the influence of the loss tangent values for the side walls.

Variation of $tan\delta$ of the tyre side wall influences the radiation amplitude but not the directivity. First influence on the radiation pattern is observed with $tan\delta = 0.5$. 5 dB are lost. Further 5 dB (dependent on the polarization and plane) are lost with $tan\delta = 1$. This is a very important remark for the further tyre modelling.

Due to very small volume and thin layer, apex and bead reinforcement were analyzed simultaneously. Fig. 3.17 shows the influence of the permittivity values for the bead reinforcement together with apex.

Fig. 3.18 shows the influence of the loss tangent values for the bead reinforcement together with apex.

It appeared that neither ϵ_r nor $tan\delta$ have any influence on the radiation pattern. This is due to the fact that these rubber parts are positioned on the rim inner side.

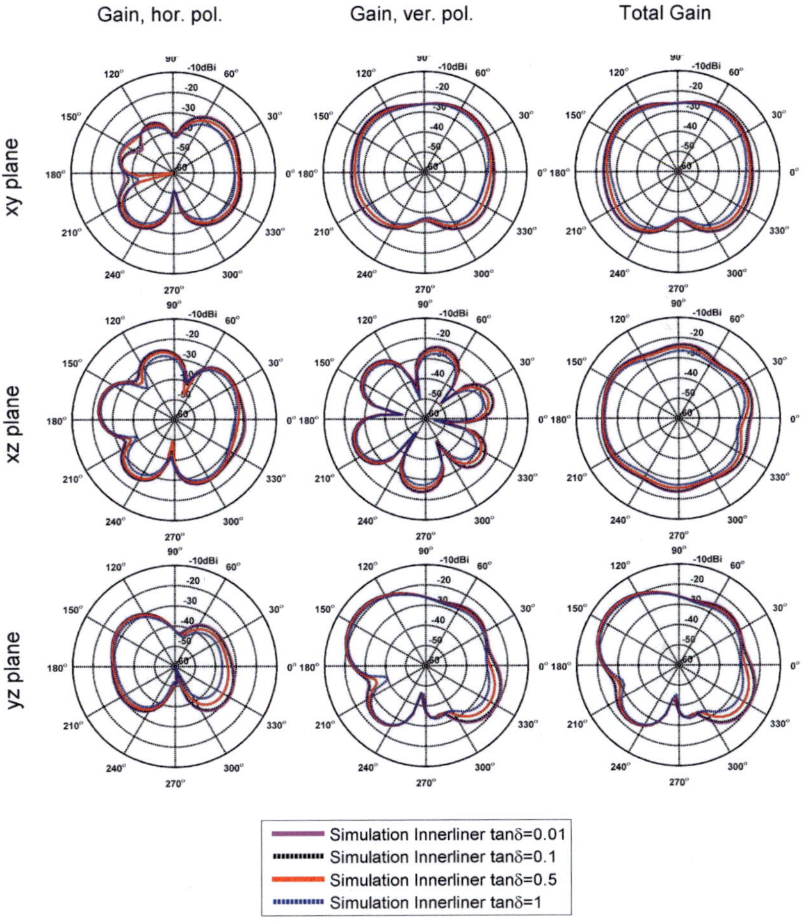

Figure 3.14: Radiation pattern simulation of 245/40 R18 tyre - variation of loss tangent values of inner liner

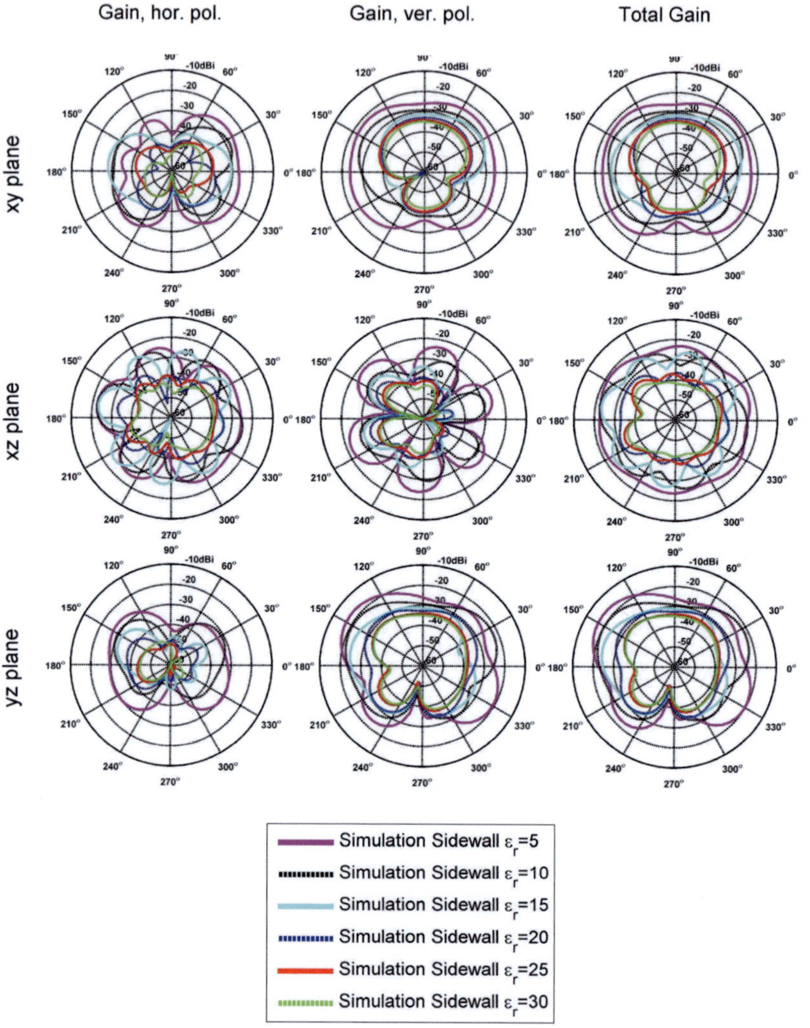

Figure 3.15: Radiation pattern simulation of 245/40 R18 tyre - variation of relative permittivity values of sidewall

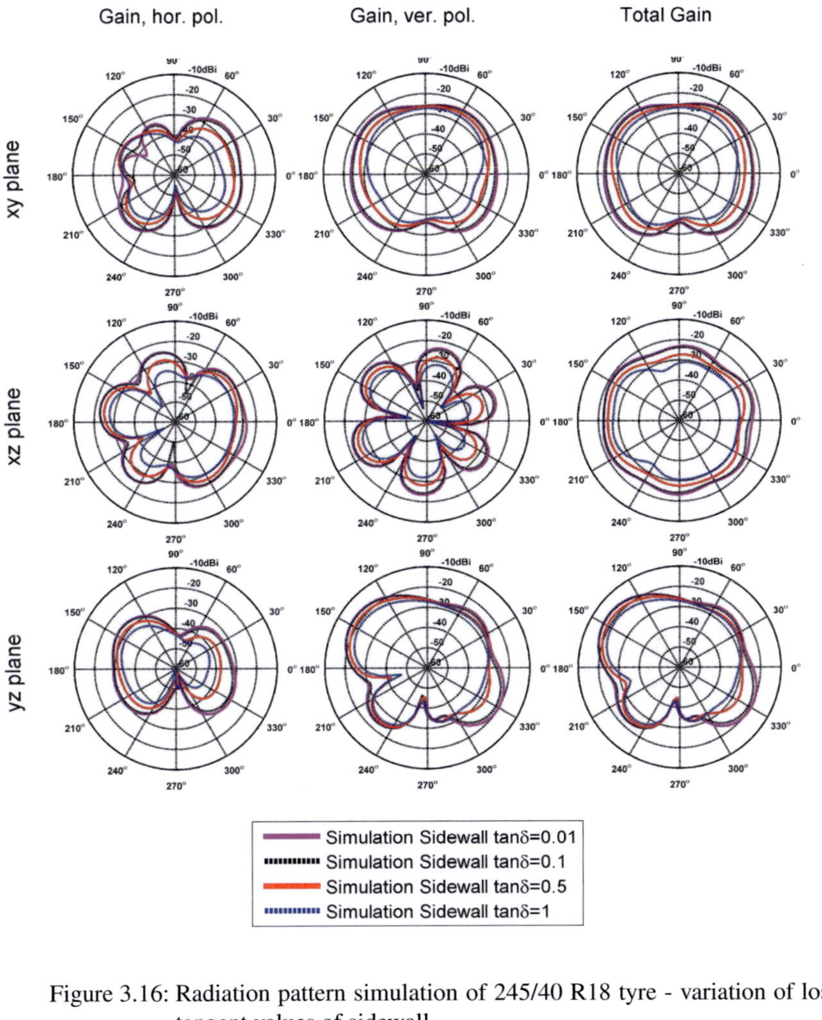

Figure 3.16: Radiation pattern simulation of 245/40 R18 tyre - variation of loss tangent values of sidewall

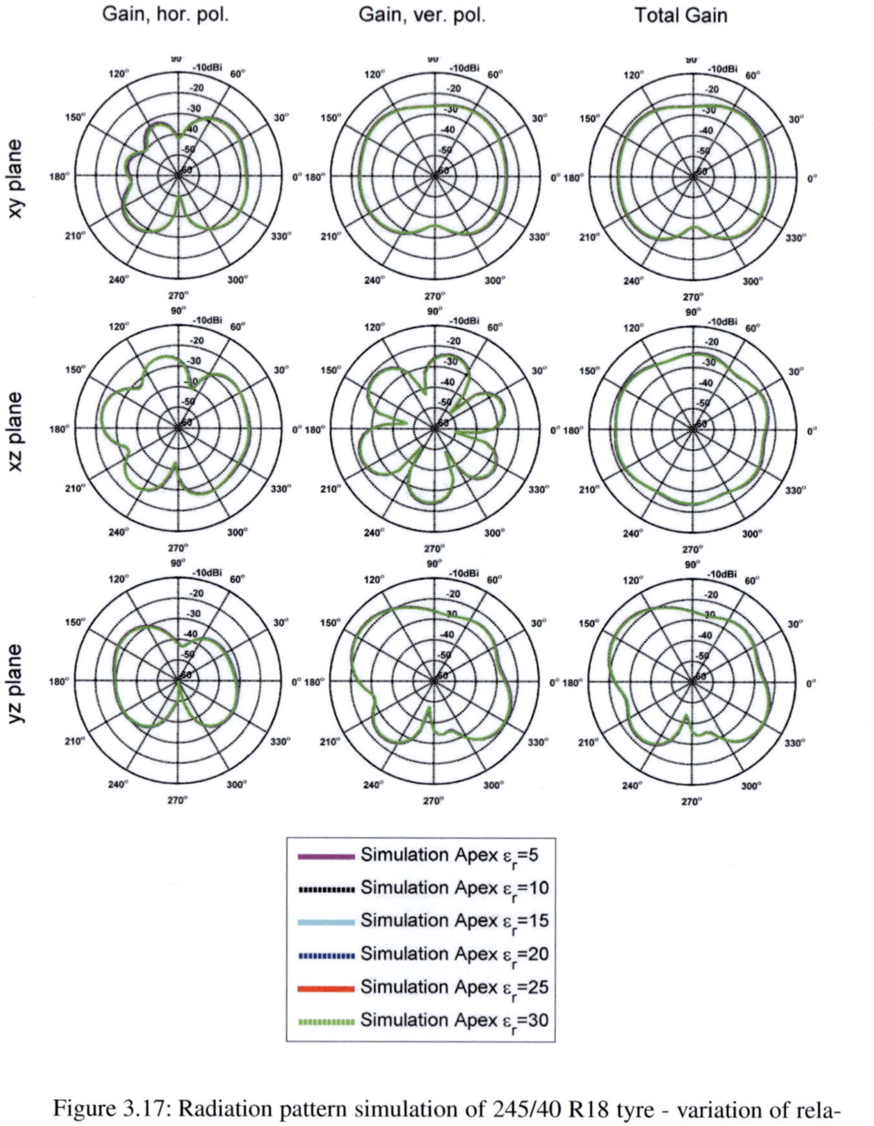

Figure 3.17: Radiation pattern simulation of 245/40 R18 tyre - variation of relative permittivity values of bead reinforcement and apex

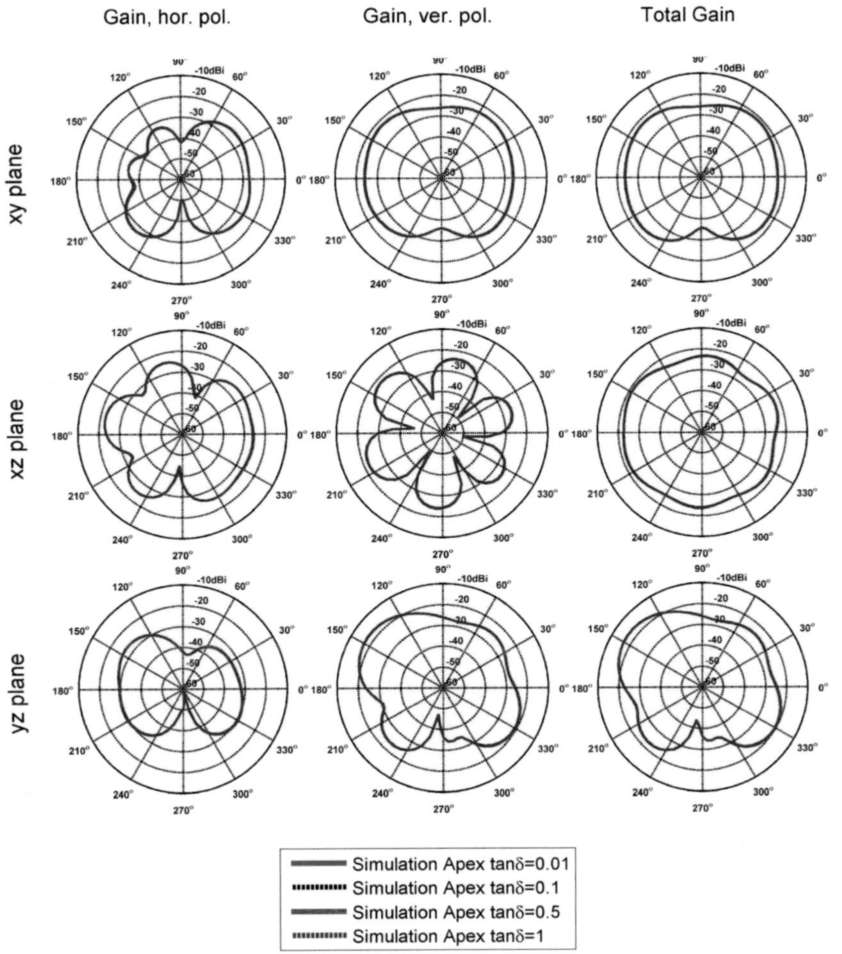

Figure 3.18: Radiation pattern simulation of 245/40 R18 tyre - variation of loss tangent values of bead reinforcement and apex

4 Material Characterization

Chapter 4 provides essential information on electric material parameter characterization. Firstly relevant parameters and therms are introduced, followed by discussion on the choice of appropriate material measurement methods especially applicable for tyre rubber (coaxial line method). Preparation of material under test samples is presented and finally followed by a discussion of the achieved results. Measured values are verified with the help of a second measurement method (loop resonator) accompanied by numerical curve fitting (HFSS simulation). All tyre types were measured and also embedding materials for WU were characterized. Detailed results for each tyre probe are gathered in Appendix D and Chapter 4 presents only the final results.

All objects and materials in the nearest proximity of an antenna play a crucial role in the radiation performance. Antennas radiate electromagnetic (EM) waves that interact with resonant and absorbing materials nearby, including the enclosure or package in which they are mounted [81]. In automotive domain this influence is of high importance. Therefore this is to be carefully considered whilst antenna design. Usually the available space for an antenna is very confined. The radiating element [82] is often very close or lies directly upon, above or aside of further vehicle parts. These may be metallic surfaces of the vehicle body, wiring harnesses, ECUs (Electronic Control Unit), plastic or as in a tyre rubber parts.

It is not only important to know the distance and exact orientation of the antenna module in respect to such objects but also electrical properties of the surroundings are of interest. Metallic surfaces will cause reflection (in order to satisfy the boundary conditions [83]) or coupling problems and rubber like materials will be the cause for damping [84], [85] or frequency shift effects because of dielectric loading. Knowledge about the material properties from electromagnetic point of view builds an essential input data for the sophisticated design and electromagnetic modelling of complex systems [86]. Of special interest for this work are the properties of tyre rubber material, which influence the functionality of TPMS.

4.1 Choice of Appropriate Dielectric Material Measurement Method

Material measurement methods differ from the application point of view and measurement requirements. Factors taken into account during measurement method choice are [96]:

- Frequency or frequency range of interest.
- Expected value of permittivity.

- Required measurement accuracy.
- Material properties (e.g., homogeneous, isotropic).
- Form of material (e.g., liquid, powder, solid, sheet).
- Sample size restrictions.
- Destructive or non-destructive method.
- Contacting or non-contacting the probe.
- Temperature range.

Appropriate measurement method was chosen upon underneath considerations:

- The frequency range of interest for material characterization during this thesis was determined by operation frequencies of the automotive systems (TPMS and RKE). All materials were measured in frequency range between 300 MHz and 3 GHz.
- It was assumed that the dielectric losses of the MUT are high.
- The measurement accuracy is of high importance for further accurate system simulation. The values received form dielectric measurements may be approved by measurement repeatability and second measurement method for certain chosen material probes. On the other hand, the verification for the measured values will be mirrored in good correlation between measured and simulated radiation patterns of different tyres (Chapter 6).
- Form of material, material properties and sample availability are very interesting topics. Apart from characterization of antenna module embedding materials, further goal of this thesis was the characterization of tyre material parameters. Rubber in all layers of a tyre had to be examined. Due to very complex tyre manufacturing process including vulcanization [97], the process, which establishes the material parameters, the samples must derive directly from manufactured tyres. Preparation of single, measurement dedicated rubber samples is not possible as this would influence the material properties. The material probes have to be chosen very carefully and accurately in order to receive samples of even surface without rills and curvature (Chapter 4.3).
- Adequate probe preparation (see point above) is also important for contacting the MUT.
- Temperature range of interest is determined by the automotive application, all probes were characterised in temperature range between -40 °C and 120 °C.

Taking into consideration the above demands, measurement method applicability was reviewed in the Table 4.1 and Table 4.2 [98], [99].

Non resonant methods are applied when a general knowledge of electromagnetic properties over a wider frequency range is asked. Resonant methods are used to get accurate knowledge of dielectric properties at single frequency point or several discrete frequencies [95].

As a proper and efficient measurement method is acquired as well as certainty and measurement repeatability, different measurement methods were combined.

After brief analysis of the method's advantages and disadvantages it was decided to apply primarily one of the non resonant methods. This in order to gain general knowledge of electromagnetic properties over a wider frequency range of MUT. The chosen method was the coaxial line method (see Chapter 4.2 a).

Measured parameter values were confirmed at few frequency points with the help of a resonant method (planar loop resonator - Chapter 4.2 b) combined with EM simulation of the samples. Accurate knowledge was gained on electric tyre properties.

	Method type			Advantages	Disadvantages
Non-Resonant	Reflection	Open-ended Coaxial probe	Accuracy	sufficiently accurate for medium-loss and high-loss media	dependent on the surface of the MUT, air gaps between MUT and coaxial probe affect the measurement
			Frequency range	broadband	—
			Repeatability	easy, no special sample preparation	—
			Sample	easy preparation, no machining of samples, non destructive	flat surface, semi-infinite thickness
			Temperature	high testing temperature possible	—
	Transmission Reflection	Open-ended Free space	Accuracy	—	affected by diffraction on the sample edge and multiple reflection between MUT and measuring antenna
			Frequency range	broadband, higher frequencies	low end limited by practical sample size
			Repeatability	easy, no special sample preparation, free access to the sample	—
			Sample	inhomogeneous materials, non-destructive / contactless meas.	large, flat and parallel faced sample needed
			Temperature	high temperature testing possible because of non-contacting	—
		Transmission line (Rectangular Waveguide; Coaxial line; Microstrip-line)	Accuracy	good accuracy for high loss materials; between coaxial probe and resonance cavity method	Limited low loss resolution (depends on sample thickness), measuring of several sample-thicknesses increase the accuracy
			Frequency range	broadband, microwave range	—
			Repeatability	—	difficult, every sample has to be prepared accurately
			Sample	placed in middle of transmission line, needs precisely fit the cross section of the transmission line	difficult preparation (only machinable solids); must fit into waveguide / destructive
			Temperature	—	—

Table 4.1: Dielectric measurement method applicability

	Method type			Advantages	Disadvantages
Resonant	Non-planar	Cavity resonator	Accuracy	more accurate than waveguide measurements	strong influenced by sample location inside the cavity
			Frequency range	—	only fixed frequency
			Repeatability	—	difficult, sample location has strong influence to accuracy
			Sample	suited for low-loss and medium-loss materials	precisely shaped small-sized samples needed
			Temperature	—	—
	Planar circuit	Microstrip resonator	Accuracy	high accuracy and sensitivity	—
			Frequency range	—	narrow or single frequency
			Repeatability	possible, samples can easy mount on the resonator	—
			Sample	non destructive measurement, small sample possible	contacted measurement, flat sample, only low-loss material
			Temperature	—	—

Table 4.2: Dielectric measurement method applicability

4.2 Description of the Chosen Dielectric Material Measurement Method

In the following Chapter the chosen methods and measurement methodology is briefly described. The overall employed material characterization procedure is depicted in the flow chart in Fig. 4.1.

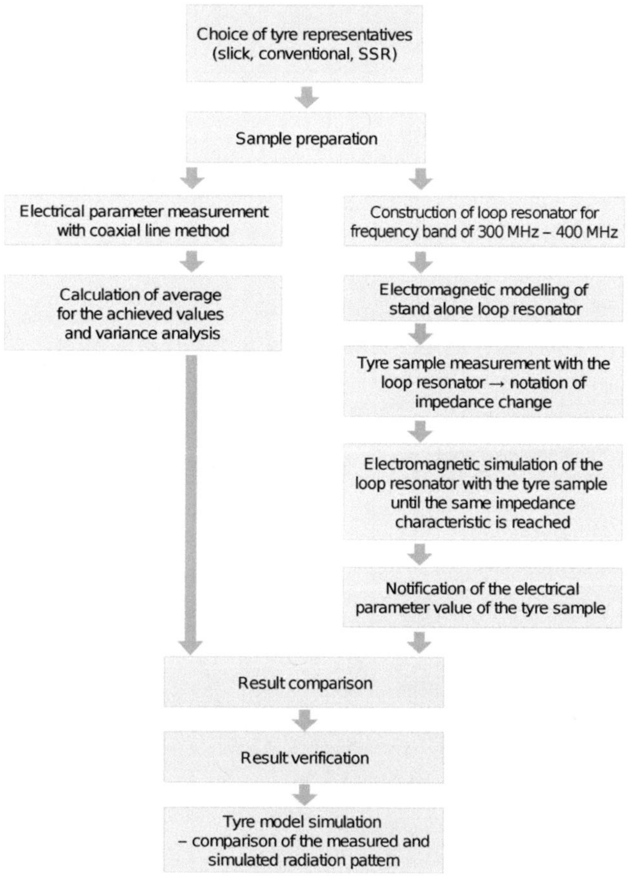

Figure 4.1: Flow chart of the employed material characterization procedure

The results of a very fast measurement method with coaxial probe are compared with the results of measurements and simulation of planar resonator method. The coaxial probe method is a very fast method and therefore all tyre samples are measured with its help. Only in order to prove the accuracy for chosen samples the planar resonator method is employed. It is proved whether the resulting values in both cases are in the same range. The final proof for material measurements is the simulation on a complete tyre system together with the WU (Chapter 6).

4.2 a Coaxial line method

Coaxial line method[1], also called open coaxial waveguide method [100], [101], belongs to non destructive measurement methods and therefore it was chosen for tyre characterization.

Fig. 4.2a presents the measurement setup. An open coaxial waveguide, called coaxial or dielectric probe (Fig. 4.2b) is pressed against surface of Material Under Test (MUT) and connected to the Vector Network Analyser (VNWA).

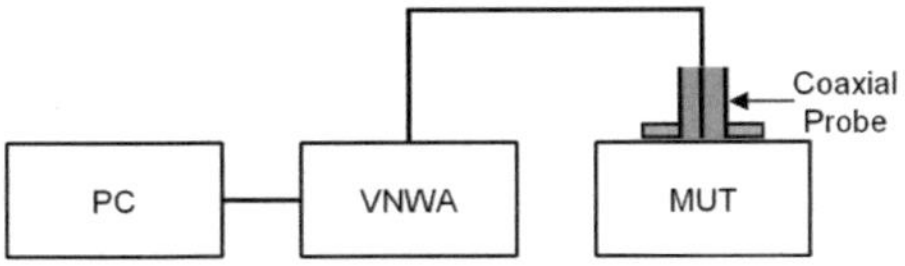

(a) Block diagram

(b) Coaxial probe

Figure 4.2: Dielectric measurement setup with coaxial line method

The dielectric properties are determined upon measured reflected power [102]. In order to calculate the dielectric parameters from the measured reflection coefficient, an equivalent circuit on an open ended coaxial line is necessary. To determine component values for the circuit an additional measurement, calibration process, with materials of well known properties is made.

For the measurements Dielectric Probe Kit from Agilent was in use with high temperature dielectric probe, which allowed measurements over, from automotive demanded, temperature range. For measurement stability, all probes of the MUT were tested in a temperature chamber (Weiss WK-111) - Fig.4.3.

[1] All results achieved with coaxial line method for tyre samples originate from cooperation activities between Continental AG and Technical University of Karlsruhe, Institute for Highest Frequency Techniques. Reference work: Report, Permittivity Measurement for Auto Tyres, J. Pontes, W. Wiesbeck, 2005

(a) Measurement equipment (b) Coaxial probe (c) MUT

Figure 4.3: Measurement equipment for coaxial line method

When applying the coaxial probe method very important are following points: assurance of cable stability, no air gaps between the probe and the coaxial probe, and adequate sample thickness. Expected behavior schematics of the coaxial line method for tyre samples of thread and inner liner is demonstrated in Fig. 4.4.

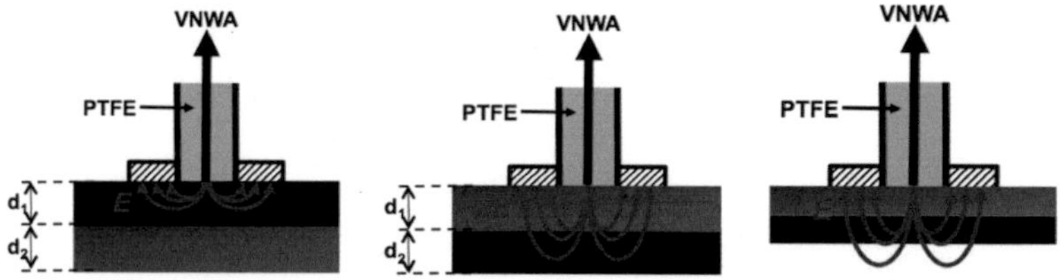

(a) Tread measurement (b) Inner liner measurement (c) Thin material probe

Figure 4.4: Schematic for coaxial line method measurement of tyre samples

For multilayer materials in the regions of higher dielectric constant (higher losses) the electric field will be bound within only one region (one layer if it is thick enough) as presented in Fig. 4.4a. For the layers with lower dielectric constant (lower losses) as it is in the case of the inner liner there is a risk (dependent on the layer thickness) that the electric field will emerge to the second layer. This phenomenon is demonstrated in Fig. 4.4b. For too thin material probes the measurement is not accurate as generated electric field is not bound within the material probe as in Fig. 4.4c.

The main advantages of the coaxial line method are that it is very convenient and the effort to prepare the right material samples is very low. Broad frequency range measurements are possible. Moreover it is a non destructive measurement method and the samples may be used once again. The only limitation is that the samples should have possibly flat surface for accurate measurements. The measurement method does not provide accurate values for liquids or semisolids but such materials are not under consideration in this thesis.

4.2 b Planar resonator method

To verify the measured dielectric properties of the MUT for few specific frequency points a planar resonator method was applied. The measurement setup is depicted in Fig. 4.5.

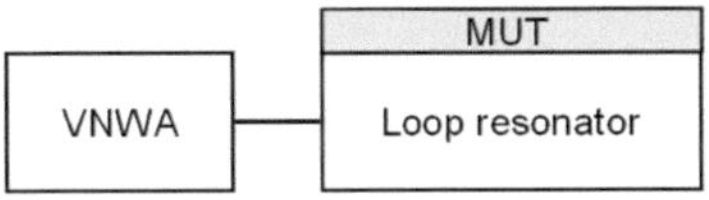

Figure 4.5: Block diagram of dielectric measurement setup with planar loop resonator

The philosophy behind this method is the comparative approach: properties of stand alone resonator are compared with the properties of the resonator when MUT lies directly on the resonating structure. The loop resonator is connected to the Vector Network Analyzer in order to measure the setup impedance.

The loop resonator was designed in order to check the correctness of the already conducted measurements for TPMS operating frequency band between 300 MHz and 400 MHz. This approach may be used however, for any desired frequency with appropriate resonator design.

Developed loop resonator, depicted in Fig. 4.6a is realized on FR4 material (ϵ_r = 4.4, $tan\delta$ = 0.02) with the size of 6 cm x 6 cm. Material thickness is 1.5 mm.

(a) Loop resonator (b) Model of the loop resonator (c) Model of the loop resonator with MUT

Figure 4.6: Resonator circuit

Parallel to the impedance measurements, the resonator structure was simulated in FEKO and HFSS. Adequate electromagnetic model of the stand alone loop resonator is presented in Fig. 4.6b and with MUT in Fig. 4.6c.

Electromagnetic simulations of the stand alone resonator were conducted in order to gain accurate model. The result of these simulations is visible in Fig. 4.7.

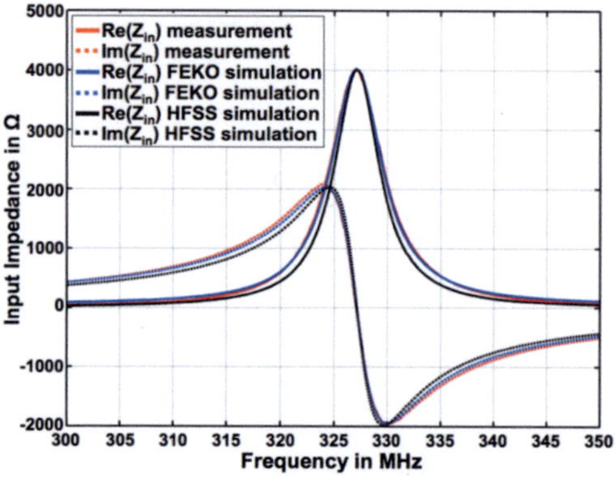

Figure 4.7: Measured and simulated input impedance of the stand alone planar loop resonator

Due to good correlation between measured and simulated values it may be stated that the resonator model both in HFSS and in FEKO is very precise.

Further, the resonator will be measured together with tyre material probe. Dependent on the material properties a certain impedance shift in the frequency range is expected as well as amplitude damping. Simulations will be conducted so long as measured and simulated values match (curves overlap) and in this way the material properties will be determined from the simulation model. The herewith obtained value is compared to the measured values with coaxial line method.

According with the above described reason, the method is considered as a comparative one: firstly stand alone planar resonator is compared to the resonator with MUT presence as well as comparison to the results of the measured values of the dielectric properties is necessary.

4.3 Preparation of the Tyre Material Samples

Exact tyre build-up is already presented in Chapter 3.2. Tyres feature a very complex structure with nine main layers (Fig. 3.3). Material properties of the layers that have influence on the RF tyre performance (Chapter 3.6) have to be measured. Due to reduced complexity and also in order to prove the measurement methodology, firstly chosen slick tyre was under examination, conventional and self supporting runflat tyres followed.

4.3.1 Slick Tyre

Slick tyres, already presented in Fig. 3.1a, do not have any kind of metallic reinforcement nor any tread profile. Therewith sample preparation is much easier than in case of conventional tyres.

Slick tyre under test and its parts are presented in Fig. 4.8. Tyre tread and side wall region was analyzed separately. Samples of tyre tread - marked as regions 1a, 1b and 2a, 2b are prepared (Fig. 4.8c) for material characterization.

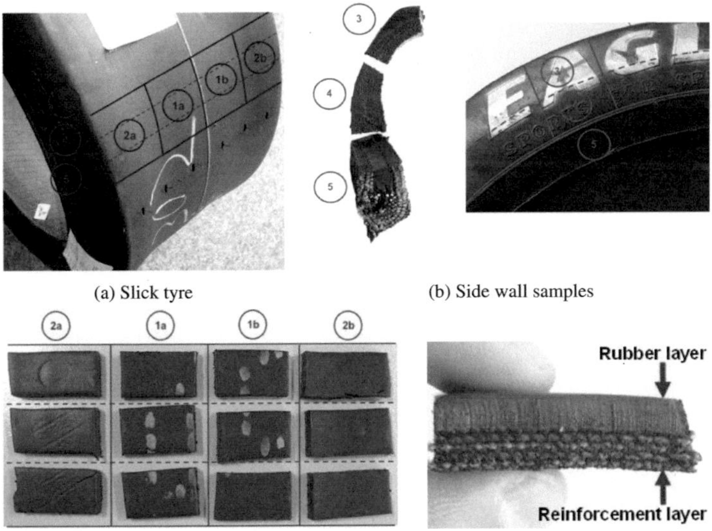

(a) Slick tyre (b) Side wall samples

(c) Tread samples (d) Cross section of the tread layer

Figure 4.8: Slick tyre - samples for measurement of electrical material properties

Tyre tread consists of sections: upper tread part out of rubber and inner liner rubber reinforced with fibre - Fig. 4.8d. Such constellation eases the probe preparation and the measurements. Slick tyre side wall may be divided into three different regions Fig. 4.8b.

The lowest part (marked as region 5) has got the most complicated structure. It consists of the metallic bead core and apex is formed by nylon reinforcement with rubber around it. A way to overcome sample preparation and separation of these regions is to assume that the rubber surrounding the bead is the same as for the rest of the tyre wall and to omit the fibre reinforcement. Finally comparison of the tyre and sensor measurement with the simulation will confirm the rightness of this simplifying assumption. The reason for this assumption is also the fact, that it is not expected that this tyre region will influence the radiation properties very much as this tyre part sits firmly in the rim groove (what was already confirmed by investigation from Fig. 3.17 and Fig. 3.18). Therewith only dielectric properties of the middle and upper segment (region 3 and 4 from Fig. 4.8b) are to be identified. Several difficulties are to overcome in this measurement: the tyre curvature and the layer thickness, which changes along the side wall.

4.3.2 Conventional Tyre

Due to more complex structure conventional tyres call for more accurate analysis of the tyre mechanical structure and very careful preparation of the measurement samples. The tyre that was chosen as representative for conventional tyres is Pirelli P Zero Rosso (245/40 R18). Similarly as in the case of slick tyres, also tread and side wall structure was analysed separately. The tyre under test is depicted in Fig. 4.9.

Figure 4.9: Conventional tyre structure

The differences that make this tyre more complicated to prepare samples are: the profiled tread and the metallic reinforcement between the tread and the inner tyre layer. Two wire layers may be recognized in the tread (Fig. 4.9, see cross section).

In order to gain more knowledge about the metallic reinforcements, the chosen tyre underwent an X-ray. Two different metallic reinforcements may be recognized. One extends in the tread (Fig. 4.10a), the second one is the bead core (Fig. 4.10c). Further the tyre is free from any other kind of metallic elements. The bead core may be at the moment neglected, whereas the tread reinforcement demands more attention.

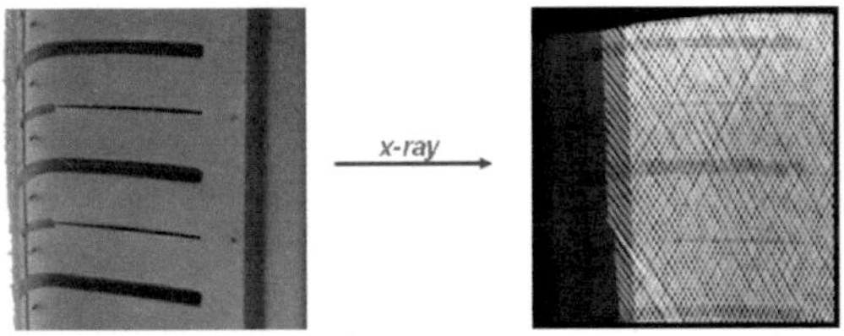

(a) Tread of conventional tyre

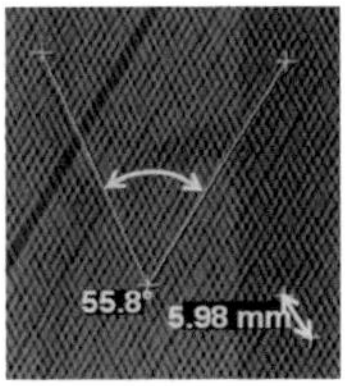

(b) Tread section

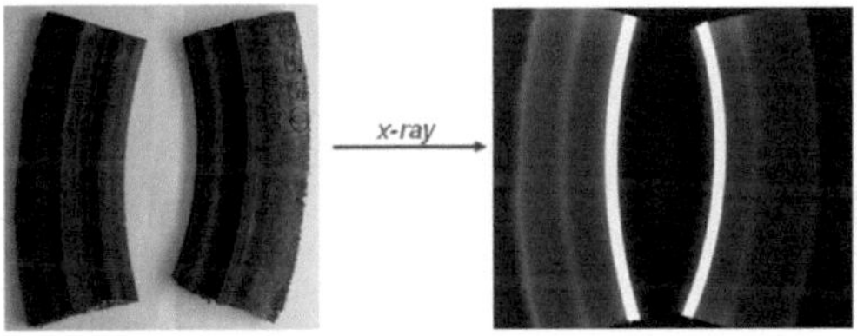

(c) Side wall of conventional tyre

Figure 4.10: Conventional tyre X - ray

It may be easily recognized that wire belts run very close to each other and cover the whole tread surface. The distance between the wires was measured and with 6 wires per 5.98 mm distance, results in gaps between the wires of less than 1 mm. The wires run in two planes, diagonally to each other, crossing at 55.8 ° (Fig. 4.10b).

For the measurement purposes the tread was divided into 5 sections (Fig. 4.11a). From each section samples of minimum 2 cm x 2 cm are obtained. Tyre rubber is treated so that it was possible to obtain probes of pure tread and inner liner rubber (Fig. 4.11b).

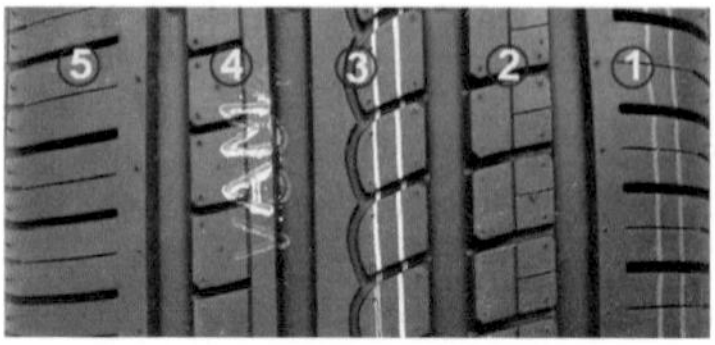

(a) Tread

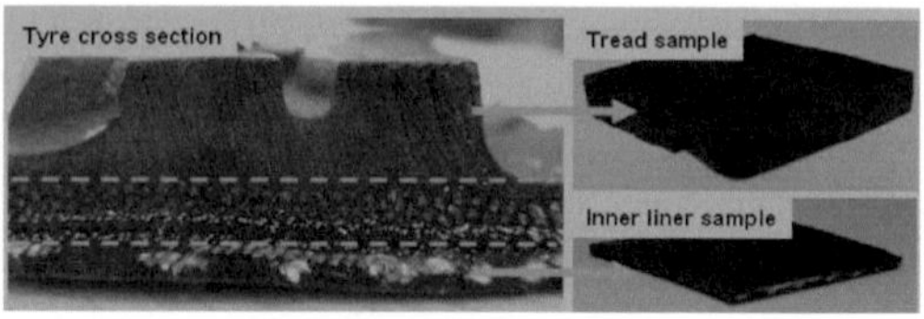

(b) Separated rubber samples

Figure 4.11: Conventional tyre - samples for measurement of electrical material properties

4.3.3 Self Supporting Runflat Tyre

Mechanical structure of Self Supporting Runflat tyres was already discussed in Chapter 3.1. This tyre type features, in comparison to conventional tyres, an extra rubber layer in the side wall (Fig. 3.2a). Dependent on the manufacturer, the additional supporting rubber layer is positioned either between the inner liner and the side wall outer rubber or is added from the tyre inside so that the inner liner turns to middle layer. Cross section of tyre under test Continental (Sport Contact 3) is depicted in Fig. 4.12b.

(a) Tread

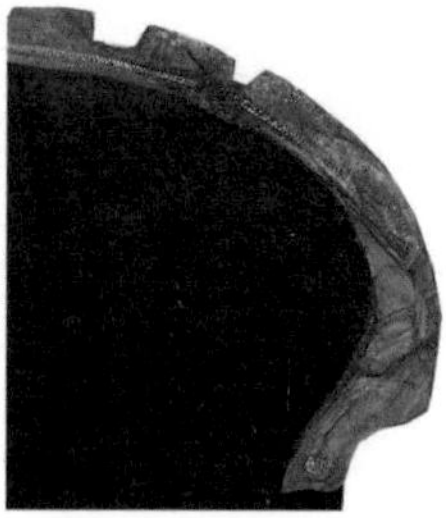

(b) Cross section of tread and side wall

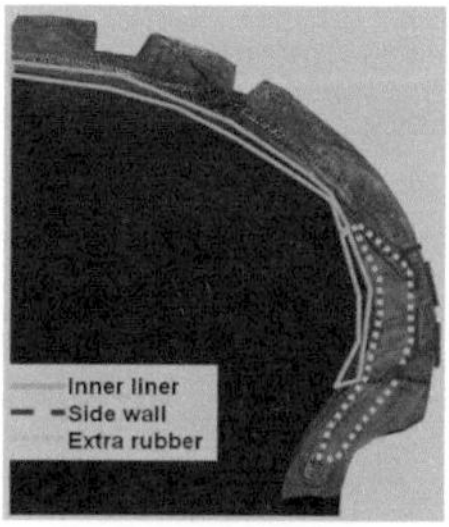

(c) Indicated rubber parts for sample preparation

Figure 4.12: Self Supporting Runflat tyre parts

Samples for the tread area are gained in similar manner as in the case of conventional tyres - both tyre types feature here the same build-up. Due to strong tyre curvature only samples out of three distinct thread segments are obtained (Fig. 4.12a).

For the side wall region three different layers are recognizable and therefore sample of inner liner, side wall outer part and extra rubber, as indicated in Fig. 4.12c, are to be prepared and analysed.

4.3.4 Additional Measurement Remarks

To assure measurement repeatability, equal pressure (of 20 N/cm^2) is applied with the help of an analogue pressure gauge on each rubber probe during consequent measurements.

For each tyre region numerous samples are measured. An average of all conducted measurements is build. All figures (from Fig. 4.13 to Fig. 4.22) feature obtained average values (solid line) and variance of the measured results (vertical lines). This is valid for all subsequent material measurements with the dielectric probe of all tyres.

Firstly electric material properties of chosen representative tyres are characterized only at the room temperature (25°C). All measurements are conducted in the frequency range between 300 MHz and 3 GHz.

4.4 Slick Tyre Material Characterization

All conducted measurements for slick tyre are gathered within the three following figures. Respectively Fig. 4.13, Fig. 4.14, Fig. 4.15 for tyre tread, side wall and inner liner.

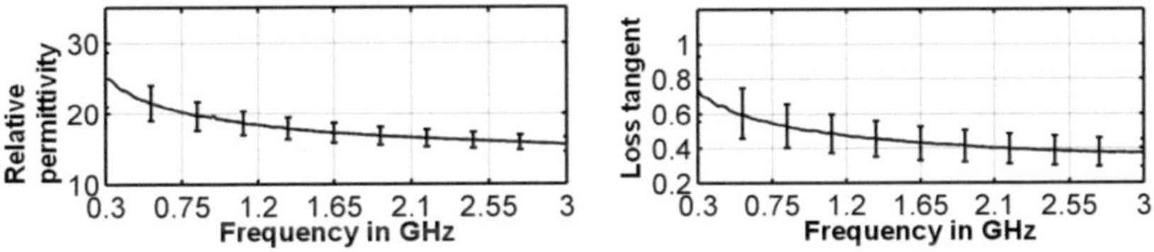

(a) Relative permittivity $\epsilon_r(f)$ (b) Loss tangent $tan\delta(f)$

Figure 4.13: Measured electrical parameters of slick tyre tread

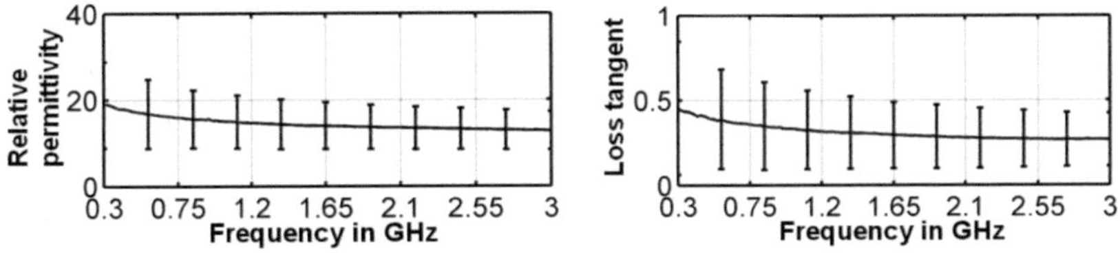

(a) Relative permittivity $\epsilon_r(f)$ (b) Loss tangent $tan\delta(f)$

Figure 4.14: Measured electrical parameters of slick tyre side wall

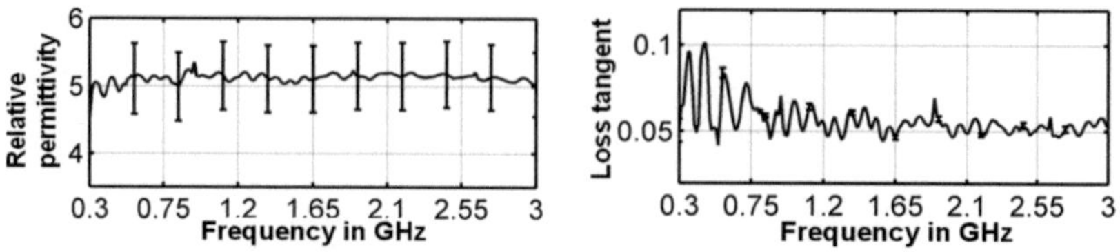

(a) Relative permittivity $\epsilon_r(f)$ (b) Loss tangent $tan\delta(f)$

Figure 4.15: Measured electrical parameters of slick tyre inner liner

For lower frequencies the values of relative permittivity and loss tangent show higher values in the tread and side wall region. Relative permittivity of inner liner is more stable over the whole frequency range.

The measurements confirm the expectations for the tread and side wall region, therewith the outer tyre layers to feature higher measured values as the inner liner of the tyre. This is due to relatively thin inner liner rubber layer and the influence of the fibre reinforcement.

Values of relative permittivity as well as loss tangent that are used for the tyre simulations in further Chapter 6 are gathered in Table 4.3. The displayed average values for side wall and inner layer were calculated with respect to the thickness of each of these rubber layers.

Variable	Tread	Side wall	Inner liner	Side wall and inner liner average
Relative permittivity	23	17.8	5.12	6.9
Loss tangent	0.65	0.41	0.09	0.24

Table 4.3: Measured material parameters of slick tyre at 433.92 MHz at room temperature

4.5 Conventional Tyre Material Characterization

As already mentioned (Chapter 4.3.2) conventional tyres feature much more complex structure. Detailed measurement results for numerous samples of the chosen tyre are to be found in Appendix D. Following only average values of all measured samples are presented. Respectively Fig. 4.16, Fig. 4.17, Fig. 4.18 for tyre tread, side wall and inner liner. Resulting from the above measurement, values used in the simulation model for frequency of 433.92 MHz are listed in Table 4.4.

It may be observed that the values of relative permittivity for tread and side wall are much lower for measured conventional tyre than for slick tyre. Inner liner values of conventional tyres with values of approximately 10 at the beginning of the considered frequency range are higher than for slick tyre.

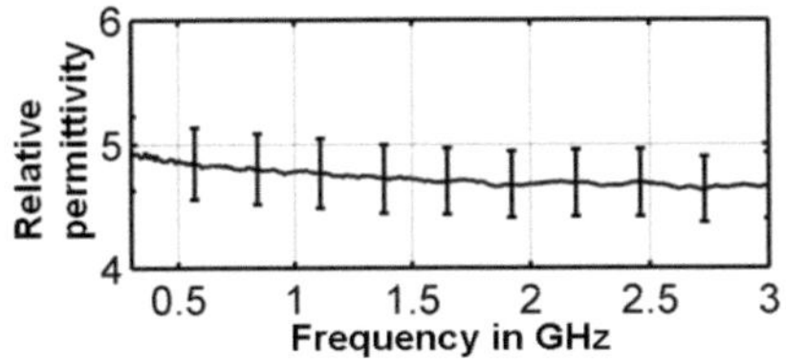

(a) Relative permittivity $\epsilon_r(f)$

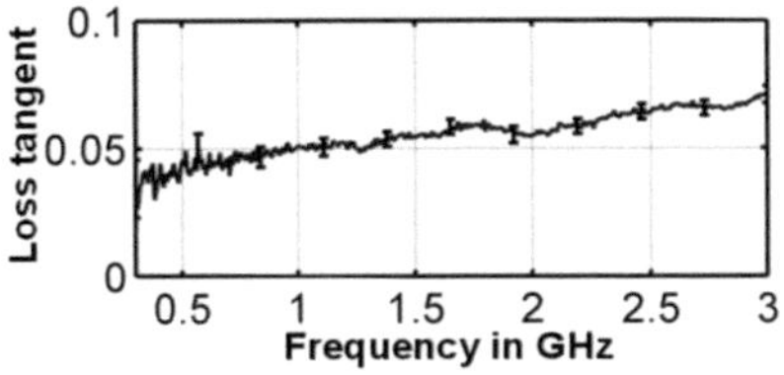

(b) Loss tangent $tan\delta(f)$

Figure 4.16: Measured electrical parameters of conventional tyre tread

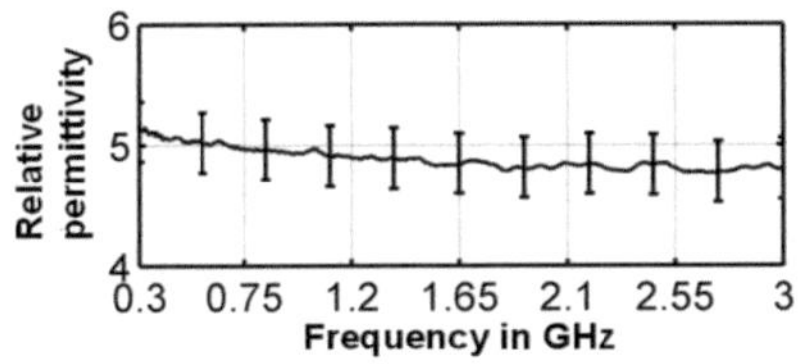

(a) Relative permittivity $\epsilon_r(f)$

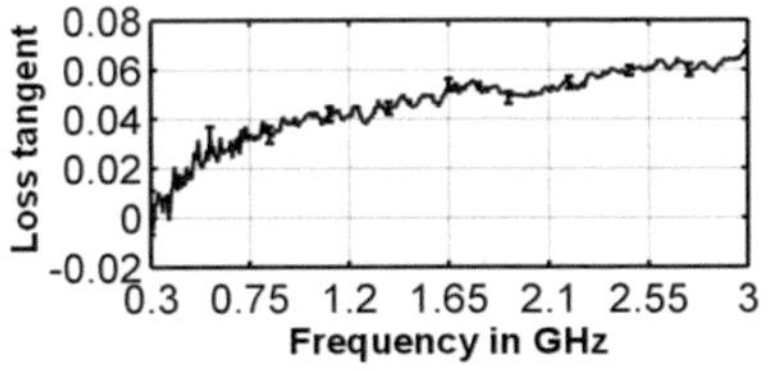

(b) Loss tangent $tan\delta(f)$

Figure 4.17: Measured electrical parameters of conventional tyre side wall

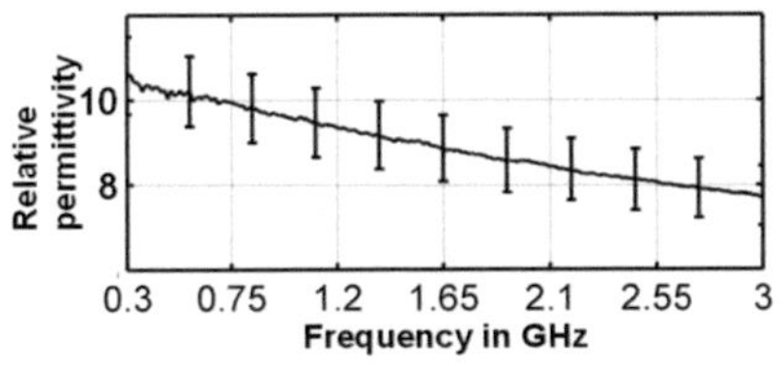

(a) Relative permittivity $\epsilon_r(f)$

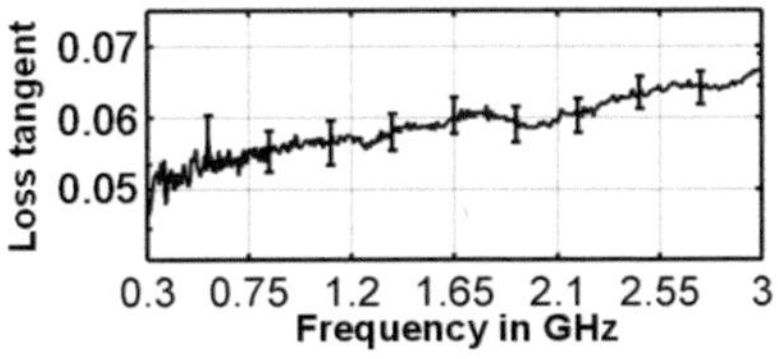

(b) Loss tangent $tan\delta(f)$

Figure 4.18: Measured electrical parameters of runflat tyre inner liner

Variable	Tread	Side wall	Inner liner
Relative permittivity	5.1	5.1	10
Loss tangent	0.05	0.05	0.05

Table 4.4: Measured material parameters of conventional tyre at 433.92 MHz at room temperature

4.6 Self Supporting Runflat Tyre Material Characterization

In case of SSR tyres characteristic values of electric parameters for tread, side wall, inner liner and additionally extra rubber were measured. In Fig. 4.19, Fig. 4.20, Fig. 4.21 and Fig. 4.22 the results are depicted for above listed tyre parts respectively. Resulting values used in the simulation model for frequency of 433.92 MHz are listed in Table 4.5.

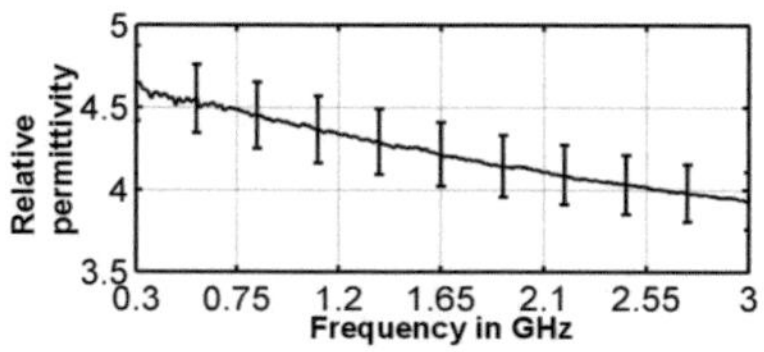

(a) Relative permittivity $\epsilon_r(f)$

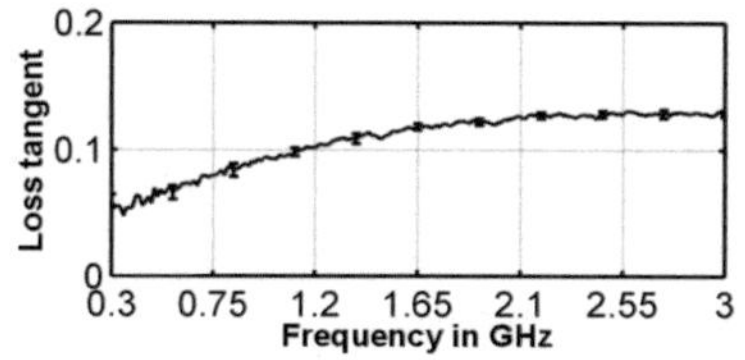

(b) Loss tangent $tan\delta(f)$

Figure 4.19: Measured electrical parameters of runflat tyre tread

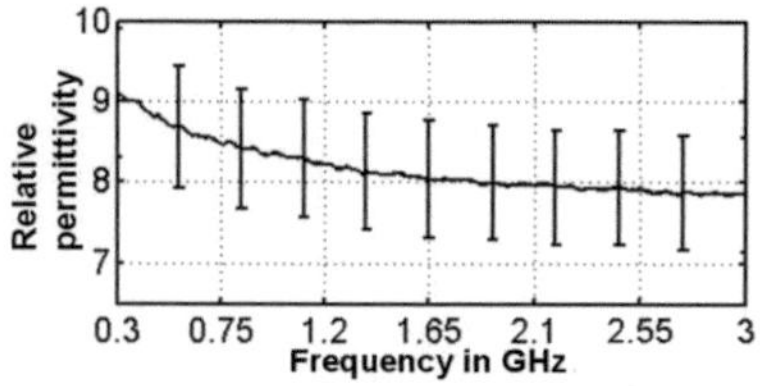

(a) Relative permittivity $\epsilon_r(f)$

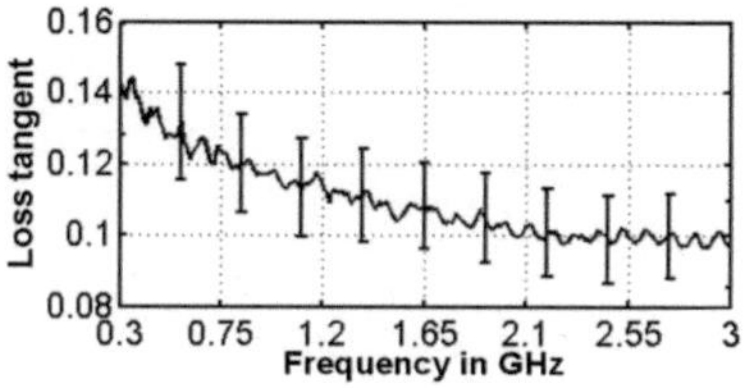

(b) Loss tangent $tan\delta(f)$

Figure 4.20: Measured electrical parameters of runflat tyre side wall

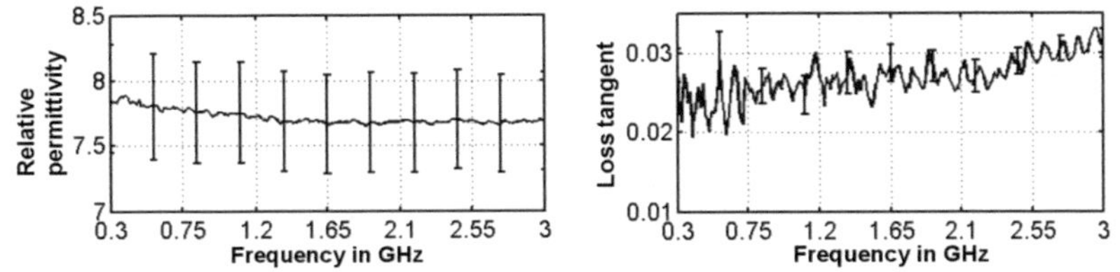

(a) Relative permittivity $\epsilon_r(f)$ (b) Loss tangent $tan\delta(f)$

Figure 4.21: Measured electrical parameters of runflat tyre inner liner

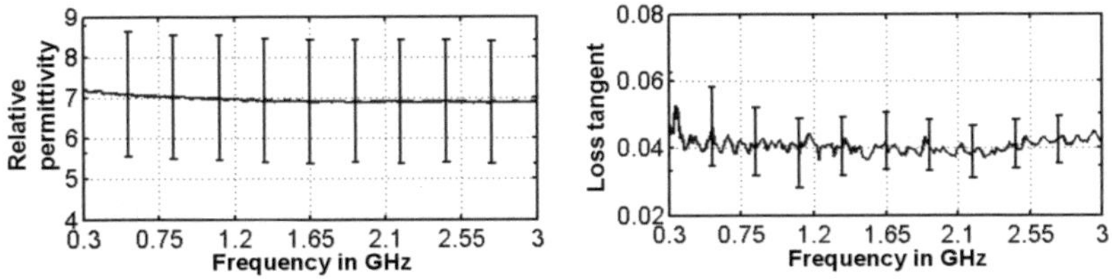

(a) Relative permittivity $\epsilon_r(f)$ (b) Loss tangent $tan\delta(f)$

Figure 4.22: Measured electrical parameters of runflat tyre extra rubber

Variable	Tread	Side wall	Inner liner	Extra rubber
Relative permittivity	4.5	8 - 9	7	7
Loss tangent	0.06	0.13	0.02	0.04

Table 4.5: Measured material parameters of SSR tyre at 433.92 MHz at room temperature

4.7 Verification of the Measured Values

Tyre electrical values measured with the coaxial line method were approved by the planar resonator method. The comparison is shown only for the chosen tyre samples. This was tread and inner liner of the conventional tyre.

The results for tread are depicted in Fig. 4.23 and for the inner liner in Fig. 4.24. In both cases firstly measurement was done of the stand alone resonator and with MUT.

It may be observed that due to rubber influence the resonant frequency is shifted to the lower band (Fig. 4.23a, Fig. 4.24a). Materials with higher relative permittivity cause larger frequency shift.

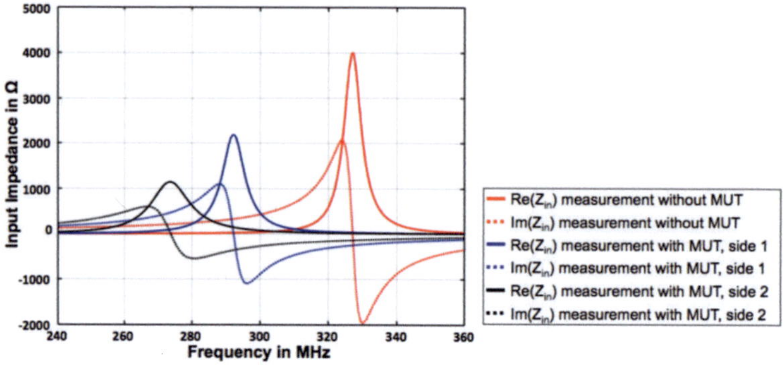

(a) Measurement with and without MUT

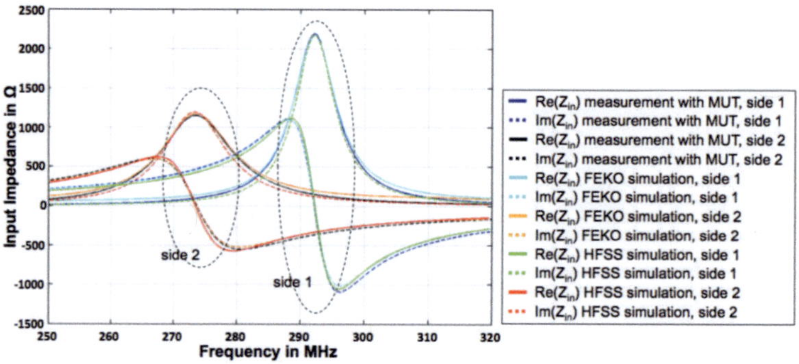

(b) Comparison of measurement and simulation with MUT

Figure 4.23: Measured and simulated input impedance of loop resonator with and without MUT - tread of conventional tyre

Tyre tread sample was laid on the loop resonator structure twice - side 1 (tyre outer surface - tread directly on the resonator structure) and side 2 (sample surface bordering with tyre reinforcement). For both cases simulations were done with appropriate material parameters so that the simulated impedance matched with the measured values (Fig. 4.23b).

Values of electrical parameters for MUT were for side 1: ϵ_r = 4.3, $tan\delta$ = 0.05 and for side 2: ϵ_r = 7.1, $tan\delta$ = 0.1 simulated with HFSS. Simulated with FEKO for side 1: ϵ_r = 4.6, $tan\delta$ = 0.05 and for side 2: ϵ_r = 7.3 and $tan\delta$ = 0.11. Higher vales of relative permittivity in the case of side 2 may be explained with the influence of the inner liner, (measured ϵ_r = 10 for inner liner, Fig. 4.18) - at the region with metallic reinforcement both rubber types from tread and inner liner are mixed.

Inner liner sample was measured and simulated only once. Both measured and simulated in FEKO and HFSS input impedance with MUT match very well - Fig. 4.24b.

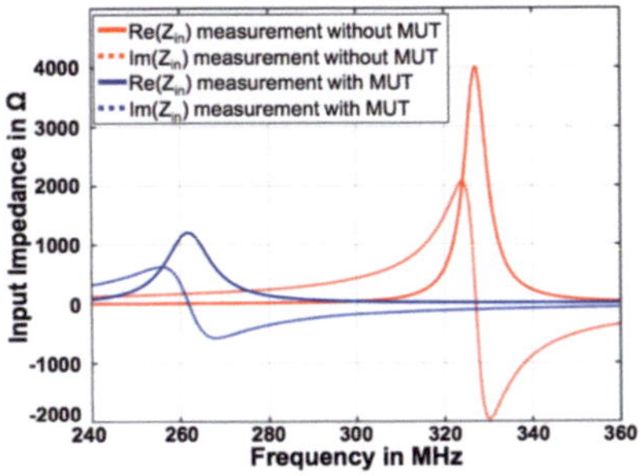

(a) Measurement with and without MUT

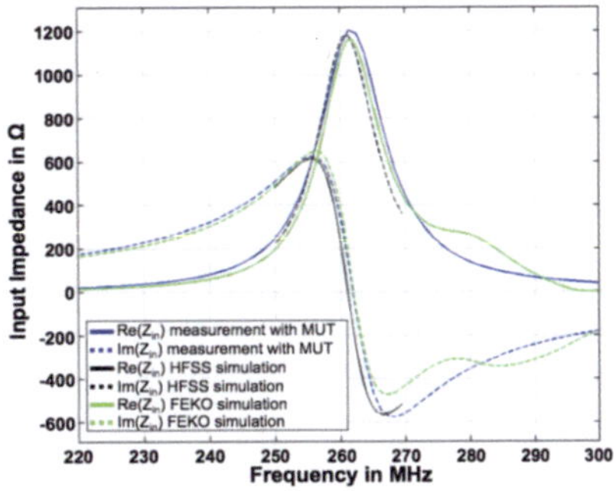

(b) Comparison of measurement and simulation with MUT

Figure 4.24: Measured and simulated input impedance of loop resonator with and without MUT - inner liner of conventional tyre

The used material values for the above material were in FEKO $\epsilon_r = 10.8$, $tan\delta = 0.08$ and in HFSS $\epsilon_r = 10.2$, $tan\delta = 0.06$.

Both FEKO and HFSS models are in very good correlation. Analysing the material parameters (which result out of the above comparative loop resonator method and values resulting from measurements with coaxial probe) for conventional tyre (Table 4.4), it may be stated that all values are a very good agreement. With this experiment already the certainty of the measured tyre parameter values was achieved - further confirmation of these values is the agreement between measured and simulated tyre radiation pattern.

4.8 Antenna Housing and Embedding Material Characterization

Antenna housing and embedding material RF properties play a crucial role while designing the antenna and determining its performance independent whether this is the WU antenna or antenna for the receiver module. As presented further in Chapter 5.3.3 material utilized for the WU housing and embedding material, were accordingly PA66-GF30 (glass fibre reinforced polyamid) and polyurethane. The measurement of the electrical properties was conducted with the coaxial transmission line method (Chapter 4.2).

The materials are characterized over the frequency range from 300 MHz up to 500 MHz and temperatures of 25 °C and 105 °C. It can be observed that both relative permittivity and loss factor of both the materials increase at higher temperature. Fig. 4.25a and Fig. 4.25b present the measurement results of the relative permittivity and loss tangent respectively. The exact parameter values for the frequency of interest (433.92 MHz, TPMS) are given in Table 4.6. These values provide a reference for all further investigations.

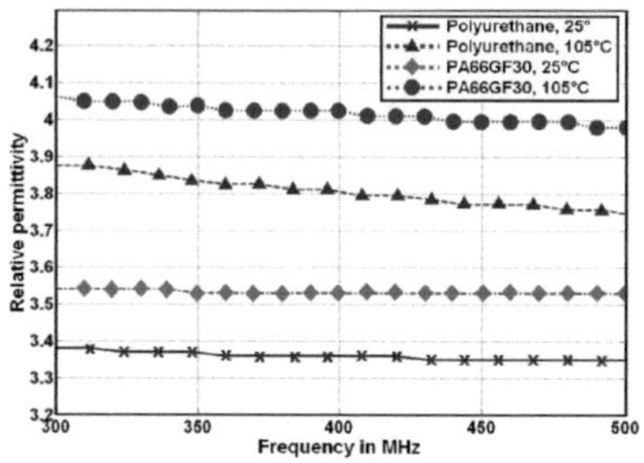

(a) Relative permittivity $\epsilon_r(f)$

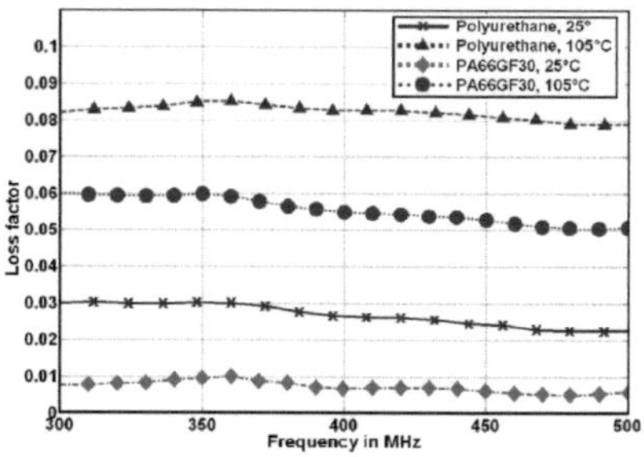

(b) Loss factor $tan\delta(f)$

Figure 4.25: Measured material parameters of the Wheel Unit housing and embedding material

Further Polyurethane was compared with Silicon as an alternative for the embedding material. The characteristic parameters over the considered frequency range and two temperature ranges are to be read from Fig. 4.26.

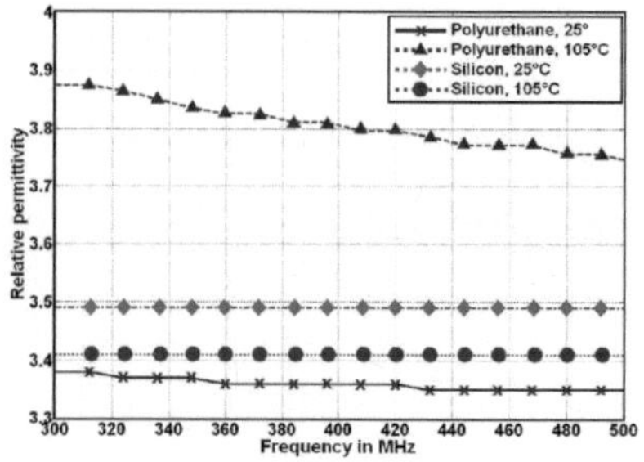

(a) Relative permittivity $\epsilon_r(f)$

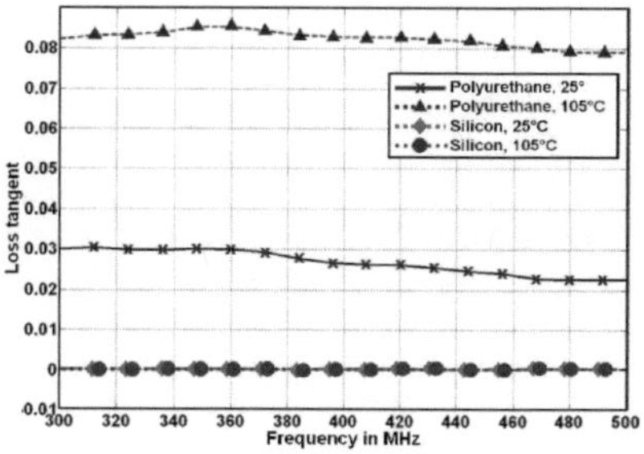

(b) Loss factor $tan\delta(f)$

Figure 4.26: Comparison of the measured material parameters of the Wheel Unit embedding materials: Polyurethane and Silicon

It may be recognized that Silicon features advantageous performance in therms of parameter stability. Both relative permittivity and loss factor are not dependent from temperature nor frequency of consideration. Moreover losses are negligible. Values for further consideration at the frequency of 433.92 MHz are gathered in Table 4.7.

Material	PA66-GF30 (housing material)		Polyurethane (embedding material)	
Temperature	25 °C	105 °C	25 °C	105 °C
Relative permittivity	3.55	4	3.36	3.8
Loss factor	0.007	0.055	0.025	0.082

Table 4.6: Measured material parameters of the Wheel Unit housing and embedding material at 433.92 MHz

Material	Polyurethane (embedding material)		Silicone (embedding material)	
Temperature	25 °C	105 °C	25 °C	105 °C
Relative permittivity	3.36	3.8	3.49	3.4
Loss factor	0.025	0.082	0	0

Table 4.7: Measured material parameters of the possible Wheel Unit embedding materials at 433.92 MHz

4.9 Conclusions on Material Characterization

Material characterization tasks incorporated not only the measurements but also an adequate sample preparation. Characteristic electrical parameters were defined for embedding materials under consideration in this work as well as for chosen representative, different tyre types. In order to assure result stability especially for tyre parts, the measurements were repeated for numerous samples, compared and average values were calculated. All measurements featured high stability. The minor and local jitter of the curves for tyre samples suggests occurring reflections form the metallic reinforcements. Chosen measured values achieved in measurements with coaxial probe were confirmed and assured by a complement investigation with loop resonator and computative numerical evaluation.

Additionally it is interesting to point out that measured figures reveal that all tyres feature low graphite content in the rubber parts. Typical relative permittivity for tyres with high graphite content lies between 1000 and 30 for the considered frequency range [103].

5 Antenna Development for Automotive Applications

Chapter 5 discusses antenna analysis for VAS and TPMS, advancement of already existing solutions, new design, development steps and presents achieved results.

Apart from substantial physical requirements put on antennas, depending on the application as desired operation frequency, bandwidth, appropriate radiation pattern, gain and efficiency, further demands have to be fulfilled. In automotive applications robustness is of very high importance. Antennas have to withstand mechanical shocks and environmental hazards. Also aesthetic, design, and streamline of the vehicle are of high importance. Therefore, there is not much space left in the vehicle for a separate visible antenna. From mechanical point of view, antennas have to be of small size, light weight and low profile. Producibility, simple manufacturing process with sustainability of certain accuracy is also an important but very often omitted point.

5.1 Small Antennas

There are numerous rules for considering an antenna to be electrically small. Independent on the frequency whether the antenna is intended to save space in GHz-range or is small compared to extremely long wavelengths [47], [48] [49].

Classifying an antenna into the group of small antennas is to be done always in relation to the operating frequency f and corresponding wavelength λ - so called electrical and not only physical size is to be determined.

A small antenna describes an electrically small radiating structure optimized and matched within a certain, limited volume. The significant criterium is the limitation of this volume. The most common definition is that the largest dimension of the small antenna is not more than one-tenth of a wavelength [50], [49]. In comparison the size of a standard antenna is quarter- or half-wavelength dependent on antenna type (monopole or dipole). Fig. 5.1 visualises the difference of size.

The radius of the sphere that comprises small antenna is five times smaller than for a standard ($\lambda/2$) antenna. The above definition constraints only the maximum antenna size and makes no distinction among the various methods used to construct electrically small antennas. There are many methods of reducing an antenna size [51], these may be divided into miniaturization via shaping (slot loading, bending, folding, space-filling curves, meander lines, fractal structures [52]) and miniaturization via material loading (dielectric, magnetic loading) [53].

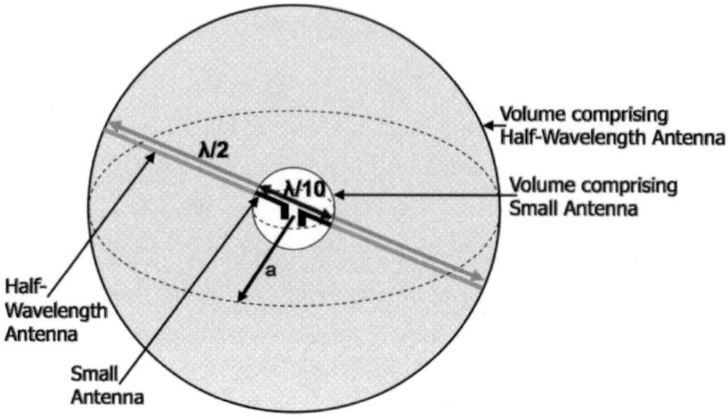

Figure 5.1: Volume comparison comprising a half-wavelength and a small antenna

The miniaturization method as well as the geometry of a small antenna are selected upon a specific application and required parameters. Apart from automotive applications the most common small antennas are the antennas implemented in portable devices. Some examples of small antennas are presented in Fig. 5.2.

(a) Wi-Fi Antenna

(b) Mobile phone antenna: GSM, UMTS

(c) Bluetooth antenna

Figure 5.2: Small antenna examples

Taking into account the antenna structure, small antennas are mostly introduced in form of short dipoles or equivalent (loaded) monopoles with ground plane, loops and patches.

5.1.1 Small Antennas in Automotive Industry

Antenna implementation in automobiles began with a single antenna for radio broadcasting. Meanwhile, due to service development, the necessity to integrate more antennas has aroused. Apart from radio AM and FM band, further services like GSM, DVB-T, UMTS,

LTE, SDARS, GPS, GLONASS, BEIDU, GALLILEO, Bluetooth, TMC, C2C, C2X, RKE, PASE and TPMS have to be taken into account. Today about twelve different antennas have to find place in one vehicle (Fig. 5.3). It is to expect that in the future this number will grow and the integration will play a very important role [12], [13].

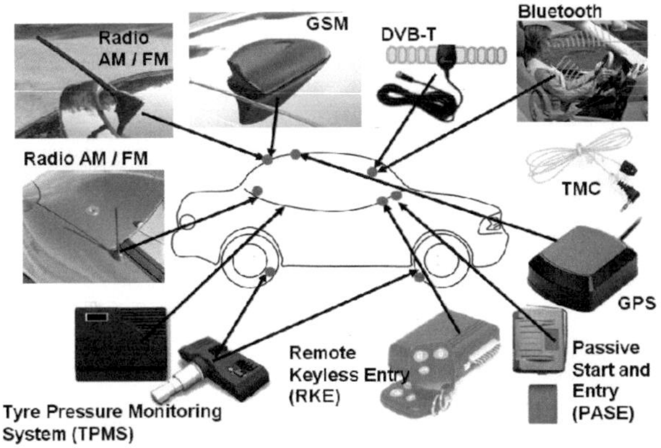

Figure 5.3: Chosen examples of automotive antennas

5.1.2 Trade Off between Antenna Small Size and the Performance

Automotive industry requires modules of small size. Should an antenna be integrated in a such module, its size implies a small antenna type. Antenna miniaturization entails certain compromises. There are many consequences that follow:

- The antenna on its own is not the only radiating element. The surrounding elements may influence and contribute to the radiation [54].
- Small antennas exhibit low radiation resistance (impedance real part Re(Z) is small) [16].
- High reactance (in general high absolute value of the impedance imaginary part Im(Z)).
- Matching network is in most cases necessary.
- When tuned to resonance a narrow bandwidth of operation results.
- Low radiation efficiency.
- It is rather hard to achieve a directive radiation pattern. Most small modules have more or less an omnidirectional, homogeneous radiation pattern. Directivity may be achieved by reflectors.

Regarding the above listed points, there are some challenges that have to be met whilst designing a small antenna. First of all the antenna has to be adapted to the module geometry and the surrounding environment where the module is going to be implemented. This constraints usually already the antenna geometry and type. It is favourable to achieve self-resonant structure. In order to make the antenna independent from slight environment changes as for example cable position, it should have maximal possible bandwidth. The matching network should not contrive high losses. It is to consider that for small antennas there is usually a small difference in the radiated power for orthogonal polarizations. Special care has to be taken regarding the highest possible antenna gain G and efficiency η. Dependent on the antenna size, there are physical limits for these parameters (Chapter 5.1.3).

5.1.3 Maximum Reachable Gain Within a Certain Volume

Every automotive module has a certain, defined size and shape. If the antenna is to be hidden, the available volume within such a module determines the maximum reachable gain for the integrated antenna. Therefore fixing the size of module is equivalent to fixing the maximum theoretically reachable gain of the antenna assuming optimal usage of the available volume. Neglecting the possible negative effects, there is a possibility to exceed the maximum theoretically reachable gain. This could be achieved for example by adequate harness positioning as the harness supports antenna radiation and represents antenna aperture extension.

The maximum reachable gain G_{max} depends on the radius of smallest sphere embracing the module a and the wavelength λ. The maximum gain limit is given by Harrington (Eq. 5.1) [55].

$$G_{max} = (ka)^2 + 2ka \tag{5.1}$$

Where k is the free space wavenumber (Eq. 5.2) and λ is the free space wavelength, a is the radius of an imaginary sphere circumscribing the maximum dimension of the antenna as shown in Fig. 5.1.

$$k = 2\pi/\lambda \tag{5.2}$$

It is important to emphasize that the Eq. 5.1 indicates an upper limit for the gain that a small antenna can achieve while still having a reasonable bandwidth. This approximation is accurate for standard antennas ($\lambda/2$), but this limit is more difficult to reach when the antenna becomes very small. In this case losses may increase drastically.

Sphere embracing the receiver (Chapter 5.2.3) and the WU (Chapter 5.3.1) to be analyzed and developed was defined with a = 7.5 cm and a = 3.5 cm respectively. Maximum reachable gain G_{max} for the given volumes is to be read from Table 5.1.

		Receiver (a= 75 mm)	Wheel Unit (a = 35 mm)
f=433.92MHz	G_{max} in dBi	3.67	3.02
f=868MHz	G_{max} in dBi	4.27	not relevant

Table 5.1: Maximum theoretical Gain G_{max} for chosen volumes and frequencies

5.2 Development of the Common Receiver for RKE and TPMS

Modern automotive applications like RKE and TPMS (Chapter 2) increase the requirements for the overall performance of the in-vehicle and in-module integrated antennas. New solutions are needed to substitute traditionally used printed and wire monopoles [56].

5.2.1 Requirements for the Common Receiver Module

The packaging of the antenna in the vehicle has an important role in terms of the antenna interaction with the car body and the intended shape of the radiation pattern. The requirements include enhanced performance in terms of range of the RKE System and received power from all wheel unit sensors for the TPMS system. These aspects are directly related to the gain, bandwidth and radiation characteristics as well as robustness and size of the antenna. The compact size is a constraint due to the available space in the car.

Summing up the mechanical and the functional requirements for a common receiver supporting two systems are the following:

- Small antenna size.
- Operation optimized for two systems (RKE and TPMS).
- Multiband operation.
- Good reception range for RKE.
- Quasi isotropic radiation pattern.
- Electromagnetic illumination of the area within the distance d around the car.
- Reliability for the communication between Wheel Unit and receiver.
- Producibility at preferably low price with high accuracy and repeatability.
- Mechanical robustness against environmental influences as vibration, humidity, etc.

5.2.2 Receiver Module Placement in the Vehicle

For TPMS systems there are generally three possible strategies for the receiver antenna placement: either there could be one antenna module placed in proximity of each wheel, the number of antenna modules could be divided by two - one module for the rear and one for the front axis, alternatively for left and right vehicle side or one antenna module that receives signals from all WU sensors Fig. 5.4. The last mentioned solution with one single receiver is preferred due to cost and assembly reasons. In this case, it is a challenge to satisfy the already listed requirements for both systems and therefore careful choice of the optimal receiver antenna position is of high importance. An optimal position may be found either following the already gathered experience or it may be examined with the help of the computer supported simulative process (developed process description in Annex C).

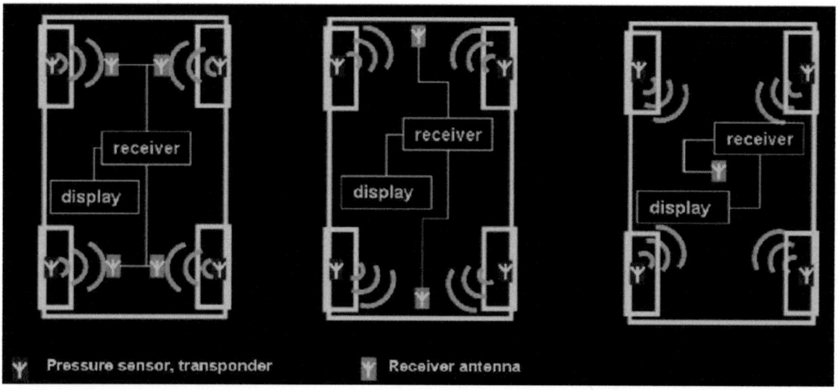

Figure 5.4: Strategies for TPMS receiver antenna placement in a vehicle

After the analysis it appeared that there are two possible preferred areas to place the receiver: one in the region of the dashboard and another one in the rear area of the vehicle under the roof. Due to the required omnidirectionality for RKE system, the receiver module was positioned as in Fig. 5.5. In order not to have any optical influence in the vehicle design the receiver was placed between the vehicle metallic roof and the headliner.

For the correct estimation of the antenna performance during the development, the placement of the module has been taken into account. Based on the position of the receiver within the car, the capacitive influence of the roof [57] on the antenna has to be considered.

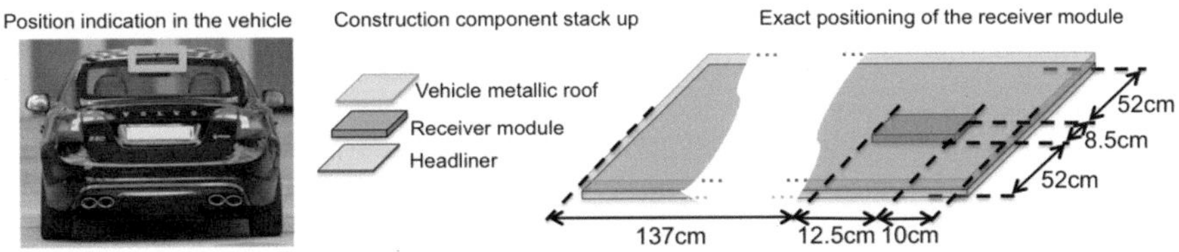

Figure 5.5: Receiver module position between the metallic roof and the headliner in the vehicle

5.2.3 Receiver Geometry and Design

As the antenna is integrated in the receiver module, its maximum size is determined by the module size. In order to fulfill all the requirements (Chapter 5.2.1) low profile Planar Inverted F-Antenna (PIFA) [58], [59] with the maximum possible dimensions was implied. PIFA structure is an attractive choice for automotive systems due to low profile, adequate radiation characteristics, and reachable wide bandwidth [60], [61], [62].

Moreover, following advantages played a key point in the decision process for a receiver antenna:

- Mechanical stability and possibility to hide the antenna in the receiver housing.
- Reduced backward radiation, which enhances antenna performance in the desired direction.
- Moderate to high gain in both vertical and horizontal polarization.
  This feature is very important due to multiple reflections on the car body and an unknown, variable position of the transmitters such as the WUs (TPMS) and the key (RKE).

5.2.3 a PIFA - Fundamental Antenna Geometry

Deducting from the name, PIFA has got a planar radiating element - the horizontal plate. This element is parallel to the ground plane. Other elements are the shortening vertical wall and the feeding source. Basic PIFA structure is depicted in Fig. 5.6.

The size and position of these basic elements determines the antenna resonant frequency f_{PIFA} (Eq. 5.3). The PIFA resonant frequency is generally approximated with Eq. 5.3 [63], [64].

$$f_{PIFA} = c_0/(4\,(l_1 + l_2)) \tag{5.3}$$

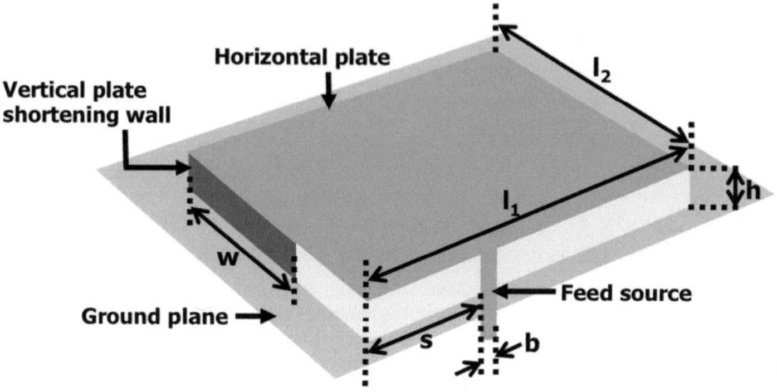

Figure 5.6: PIFA base structure

The above equation is valid for infinitesimal small grounding w ($w \approx 0$). For greater shortening wall widths, the antenna electrical size is reduced by w. Usually antenna height h is negligible as it is very small versus width and length $l_1 + l_2$, however if taken into account, it extends antenna size. Following the above rules, a more precise equation for PIFA resonant frequency is given with Eq. 5.4 [46].

$$f_{PIFA} = c_0/(4\,(l_1 + l_2 - w + h)) \tag{5.4}$$

The resonant frequency is inversely proportional to the sum of antenna lateral dimensions l_1 and l_2 as well as height h, with the broader ground strip width w resonant frequency grows.

The above structure was used as an origin to design a required dualband antenna. Firstly the antenna geometry influence on the performance and radiation properties were investigated. This in order to understand the antenna behaviour and to state the exact design rules. The theoretical investigations (as well as final simulations) were carried out with the electromagnetic simulation tool FEKO (Chapter 6).

5.2.3 b Designed Receiver Antenna Dimensions

Taking into account the maximum available space in the receiver housing, the antenna should not exceed $l_1 \leq 120\;mm$, $l_1 \leq 85\;mm$ and $h \leq 12\;mm$.

Therewith, the basic PIFA structure had to be modified in order to achieve both resonant frequencies of 433.92 MHz and 868 MHz. For this purpose according slots were introduced in order to influence current flow on the antenna structure and therewith the resonant frequency. In the literature it has been many times demonstrated that a PIFA can resonate at a second frequency with proper adaptation of slots [65], [66]. Final antenna structure is depicted in Fig. 5.7.

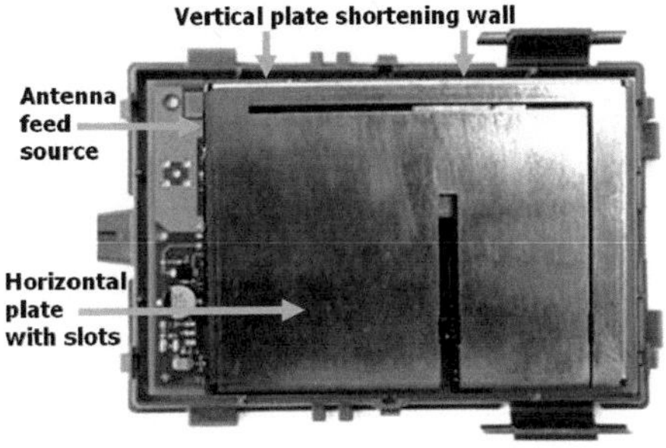

Figure 5.7: Realized PIFA structure

5.2.3 c Designed Receiver Antenna Current Distribution and Matching Network at the Required Resonant Frequency

With the application of the slot technique and influence on the current flow (simulated current distribution of a dualband antenna is depicted in Fig. 5.8a at 433.92 MHz and in Fig. 5.8b at 868 MHz) on the horizontal plate both resonant frequencies were achieved: firstly mono-band antenna operating at 433.92 MHz (Fig. 5.14a) and sequentially a dual-band antenna operating additionally at 868 MHz (Fig. 5.14b).

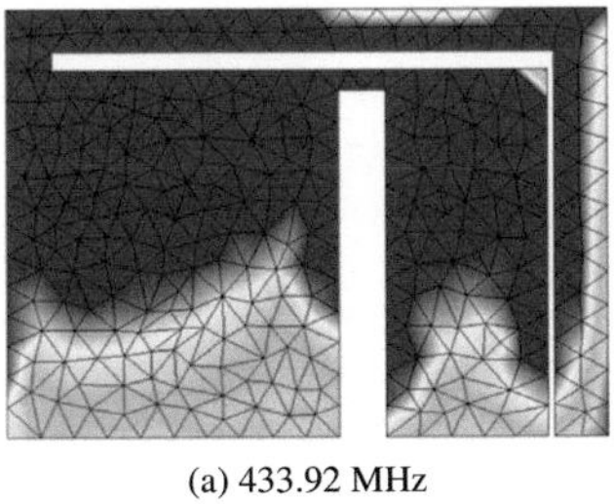

(a) 433.92 MHz

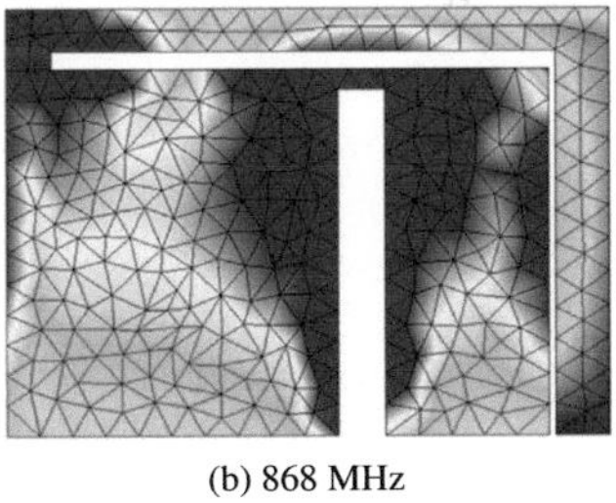

(b) 868 MHz

Figure 5.8: Simulated current distribution for the designed receiver module antenna

The optimal impedance (Fig. 5.9) of the antenna was achieved by positioning adequate shortening wall, the antenna feed and optimizing the space between the feed and the shortening wall. Some lumped components for the matching network were necessary (Fig. 5.10).

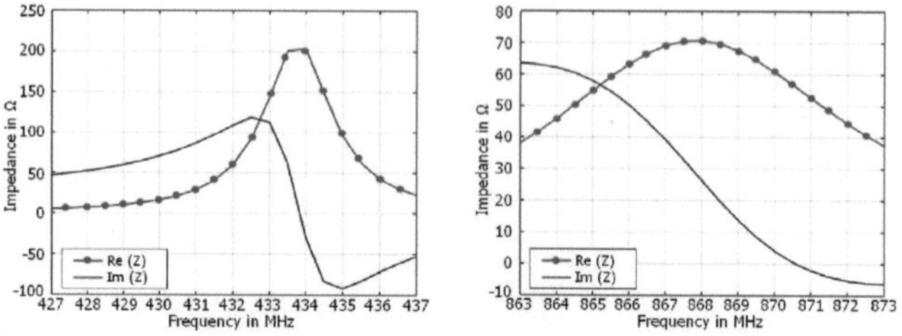

(a) 433.92 MHz (b) 868 MHz

Figure 5.9: Simulated impedance for the designed receiver module antenna

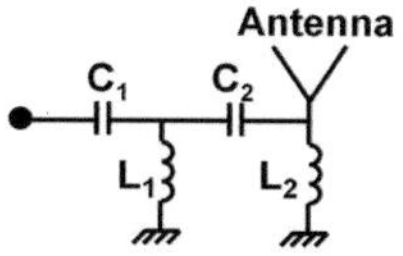

Figure 5.10: Matching network for the designed receiver module antenna: $C_1 = 4\,pF, C_2 = 2.4\,pF, L_1 = 14.6\,nH, L_2 = 33.5\,nH$

With the proposed antenna geometry and matching network components, the antenna module was simulated and subsequently measured. Simulated reflection coefficient is depicted in Fig. 5.11.

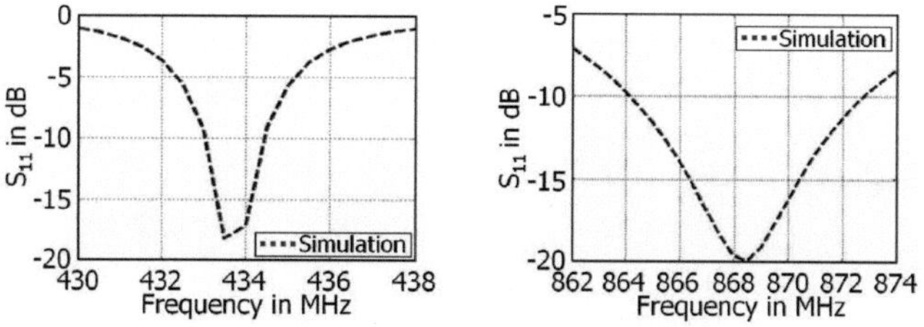

(a) 433.92 MHz (b) 868 MHz

Figure 5.11: Simulated input reflection coefficient S_{11} of the designed receiver module antenna for 50 Ω

The measurement was conducted for three probe antenna modules out of the production line. This was done in order to assure the reproducibility of the results. Fig. 5.12 depicts that the antenna is well matched at both resonant frequencies of 433.92 MHz and 868 MHz.

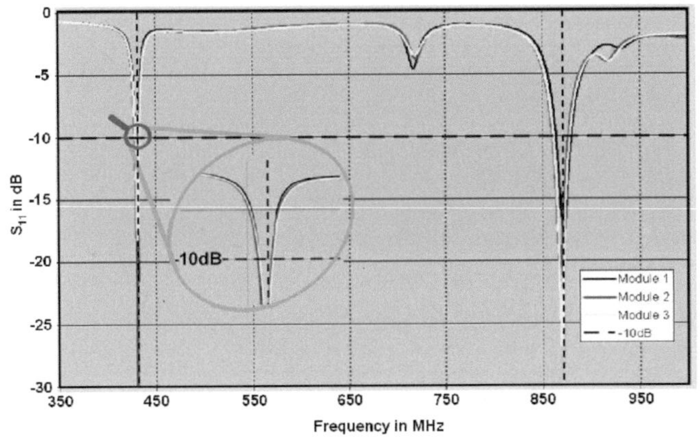

Figure 5.12: Measured input reflection coefficient S_{11} of the designed receiver module antenna (three samples: Module1, Module2 and Module3)

5.2.3 d Designed Receiver Antenna Radiation Pattern

The results for the above mentioned simulations as well as measurements were conducted in the environment that resembles real antenna environment in the vehicle. As it was decided that the most convenient place for the receiver module is underneath the roof area, for the development purposes the scenario was simplified and antenna was placed on a metallic surface as presented in Fig. 5.13a.

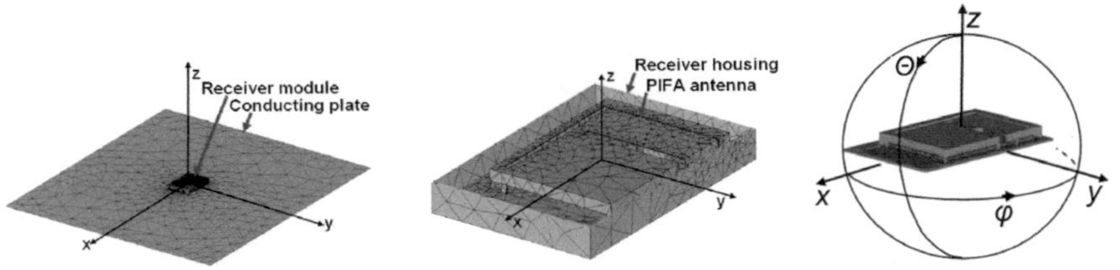

(a) Simulation model of the receiver module on the conducting plane (b) Simulation model of the antenna within the receiver model (c) Orientation of the receiver module

Figure 5.13: Receiver module orientation for radiation pattern measurement without the vehicle

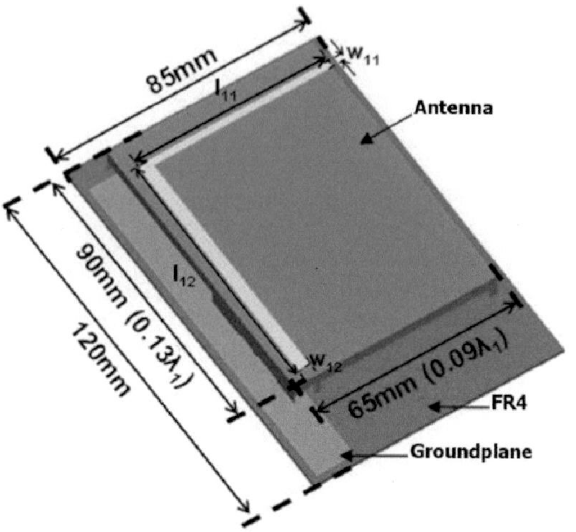

(a) Mono-band antenna for 433.92 MHz:
$l_{11} = 60.65\ mm, l_{12} = 77.85\ mm, w_{11} = 1.74\ mm,$
$w_{12} = 3.2\ mm$

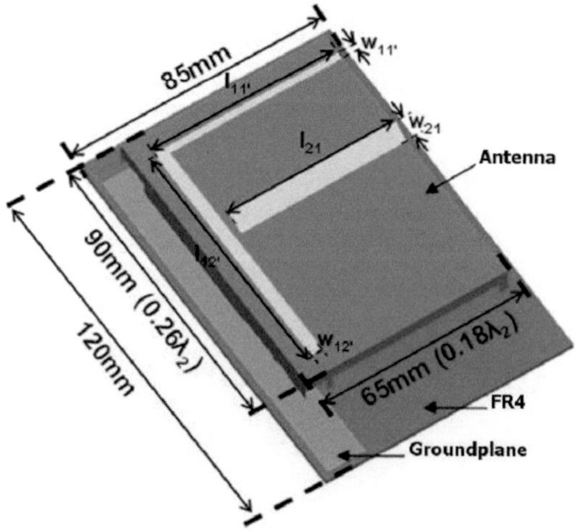

(b) Dual-band antenna for 433.92 MHz and 868 MHz:
$l_{11'} = 58.5\ mm, l_{12'} = 75.3\ mm, l_{21} = 52.5\ mm, w_{11'} = 1\ mm,$
$w_{12'} = 2.8\ mm, w_{21} = 7.1\ mm$

Figure 5.14: Geometry of the designed receiver module antenna

Fig. 5.15 and Fig. 5.16 present the radiation pattern in terms of gain for 433.92 MHz and 868 MHz, in xz and yz plane respectively - for antenna orientation see Fig 5.13c.

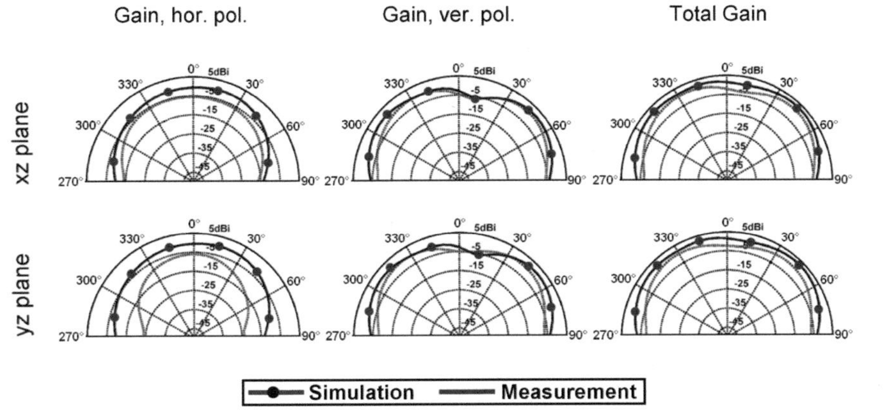

Figure 5.15: Measured and simulated radiation pattern of designed receiver module antenna for 433.92 MHz

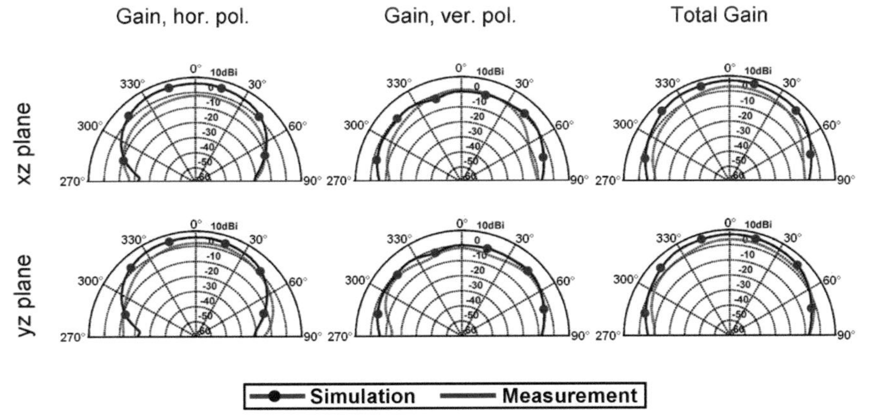

Figure 5.16: Measured and simulated radiation pattern of designed receiver module antenna for 868 MHz

For both resonant frequencies simulation and measurement results of the radiation pattern stay in good agreement. At 433.92 MHz the difference between the average of both values amounts maximal 4 dB, and at 868 MHz maximal 2 dB. Possible reasons for this discrepancy are the following:

- The measurements consider the headliner whereas the simulations not.
- Simplification of the PCB for simulation purposes.
- Uncertainty of the material parameters (especially the loss factors).
- Influence of the components on the PCB.
- Measurement accuracy.

The developed PIFA antenna for the receiver module satisfies the requirements of both systems: RKE and TPMS. Optimum geometry of the antenna and the module design together with the placement in the car was considered. The designed antenna was integrated in the receiver module without enhancing its size.

5.3 Development of the Tyre Wheel Unit Antenna

Development approach of the Wheel Unit and therewith structure of Chapter 5.3 is depicted in Fig. 5.17.

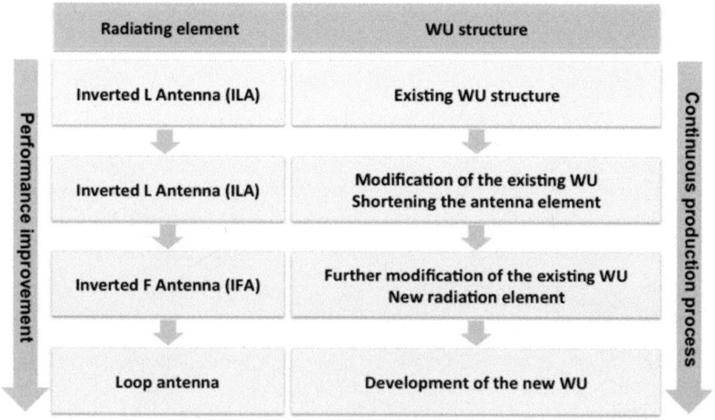

Figure 5.17: Development approach for the Wheel Unit - optimization process

The goal for the development of the Wheel Unit antenna is to analyze the existing WU, optimize the design of an internal antenna and propose new solutions for the antenna element within the given WU housing (Fig. 5.18a, Fig. 5.19a). An analysis of this WU is the first step, it is followed by modification of the existing elements in order to improve the radiation performance. After this step an exchange of the radiating element is possible, before, for the future applications, a new WU is introduced.

Just to mention other possibilities that are not discussed within this thesis, it is also possible to utilize valve as an external antenna. This solution has, however, some drawbacks as for example poor performance stability that variates in dependency on the rim profile and positioning accuracy. Looking into the future, a sensor-transponder could be integrated directly in the tyre rubber without a battery and would supply pressure and temperature data as proposed by [67]. This requires system standardization due to the fact that the sensor-transponders would have to be integrated already during the tyre manufacturing process without the knowledge on which vehicle brand and type these are later on utilized. Due to these facts most popular and flexible both for OEM and after - market TPMS solutions are still Wheel Units with an internal radiating element attached to the valve.

The development of the tyre Wheel Unit antenna requires particular attention. This is among due to the fact that a reliable communication between the in-the-vehicle positioned receiver (Chapter 5.2) and the sensor has to be guaranteed and thereby the WU transmission reliability has an important role. Additionally, volume available for the antenna structure is very confined. Therefore, an adequate electronic circuit with an appropriate antenna has to be integrated in the WU sensor. The WU sensor with the integrated antenna has to meet the following requirements (alike receiver antenna - Chapter 5.2):

- Compact size.
- Appropriate antenna gain and efficiency, fulfilling the requirements.
- Low height.
- Mechanical stability.
- Low price, which is very important in large scale industrial applications.

Moreover due to the antenna placement on the rim, the following additional requirements should be satisfied:

- Ultra low weight (due to influence on tyre rotation and wheel balance).
- Antenna parameter stability for wide range of rims and tyres.

Theoretically several different kinds of antennas could fulfill the requirements specified above. It is possible however to divide them in three categories: electric and magnetic antennas as well as combination of these two. Such antennas as a monopole, an Inverted F Antenna (IFA), a loop or a patch would be applicable.

The development and achieved advancement in the RF performance progress is discussed within the following Chapters: Monopole Antenna Analysis (Chapter 5.3.3), Inverted F Antenna Development (Chapter 5.3.4), Loop Antenna Development (Chapter 5.3.6). The assembly of the WU and requirements on RF performance are presented in Chapter 5.3.1 and Chapter 5.3.2.

5.3.1 Assembly of the Wheel Unit

During the development also the nearest antenna surrounding has to be taken into account. In a case of a WU this is primarily the rim. A fully assembled Wheel Unit together with the valve is presented in Fig. 5.18a.

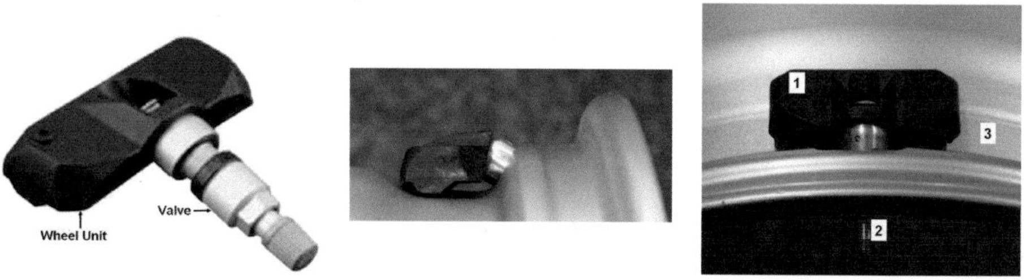

(a) Stand alone Wheel Unit

(b) Wheel Unit on the rim (side view)

(c) Wheel Unit on the rim (top view):1 - Wheel Unit, 2 - Valve, 3 - Rim

Figure 5.18: Wheel Unit

The designed WU matches any conventional rim profile and is adapted to the rim ditch (Fig. 5.18b). The exact positioning of the WU is explained in Fig. 5.18c. The WU (Fig. 5.18c-1), fastened on the valve (Fig. 5.18c-2), sits firmly on the rim (Fig. 5.18c-3).

Further the internal assembly of the WU and the components (both dielectric and metallic - Chapter 4.8) have a very important role for the antenna design as these influence the radiation performance. The WU is assembled from the following major parts depicted in Fig. 5.19: housing, Printed Circuit Board (PCB) and battery.

Additionally all above mentioned parts are moulded with an embedding material. This contributes to mechanical stability and protection against humidity. The embedding material influences however the antenna performance drastically.

Two WUs were already present in the company (Siemens VDO / Continental AG) portfolio. Each WU was equipped with a stainless steel monopole antenna of the length of 104 mm (further called short ILA) and 146 mm (further called long ILA) respectively (analyzed in Chapter 5.3.3). Utilized embedding material was polyurethane and WU housing material was PA66-GF30 (Glass Fibre reinforced Polyamid).

(a) Housing (b) PCB (c) Battery

Figure 5.19: Wheel Unit major components

5.3.2 RF Performance Requirements

The summary of the required specifications for the WU is listed in Table 5.2. The required operating frequency is 433.92 MHz with minimum bandwidth of 2 MHz. Stable performance within the temperature range according to standard automotive requirements should be guaranteed from -40 °C to 105 °C with minimum radiated power of -12 dBm and maximum loss of 3 dB over temperature range.

Parameter	Requirement
Operating frequency	433.92 MHz
Bandwidth	2 MHz at -10 dB
Temperature Range	-40 °C to 105 °C
Radiated power at 25 °C	-12 dBm
Maximum acceptable loss over the temperature range	3 dB

Table 5.2: Requirements on the Wheel Unit performance

5.3.3 Monopole Antenna Analysis

Monopole antennas in the automotive industry are typical for the AM / FM frequency ranges and have become over 50 years of deployment a traditional antenna type mounted on the roof surface, often called also whip antennas (an example was already presented in Fig. 5.3 - FM antenna).

5.3.3 a Fundamental Structure of the Monopole Antenna

The monopole antenna, shown in Fig. 5.20a, results from applying the image theory to the dipole [68]. According to this theory, if a conducting plane is placed below a single element of length L /2 ($\lambda/4$) carrying a current, then the combination of the element and its image acts identically as a dipole of length L ($\lambda/2$).

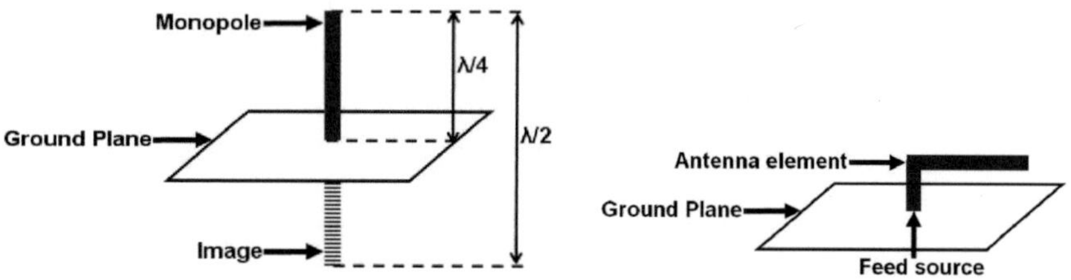

(a) Straight monopole

(b) Bent monopole:
Inverted L Antenna (ILA)

Figure 5.20: Fundamental structures of the straight and bent monopole antenna

Therefore a quarter wavelength monopole ($\lambda/4$) approximates a half wavelength ($\lambda/2$) dipole [69]. The radiation occurs only in the plane above the ground, therewith the directivity (Eq. 3.2) is doubled and the radiation resistance is halved when compared to a dipole. The typical gain (Eq. 3.1) for the quarter wavelength monopole is between 2 and 6 dB and it has a relative bandwidth of about 10 %. Its radiation resistance is 36.5 Ω and its directivity is 3.28 (5.16 dB) [70].

The desired antenna for the WU should operate at the frequency of 433.92 MHz. Considering this frequency, quarter wavelength monopole would be of an approximate length of 17.25 cm, whereas a small antenna would be of approximate length of 6.9 cm (adequately one tenth of the wavelength - Chapter 5.1). All cited and given values represent however, only a theoretical value for an antenna in the free space or over an infinite ground plane.

In practice, monopole antennas are used on finite-sized ground planes and it is so in the case of the WU. This affects the antenna properties, particularly the impedance and the radiation pattern [74]. Additionally taking given WU housing dimensions of 25.5 mm $\times$ 64.5 mm $\times$ 12 mm into account, neither a quarter wavelength, nor a small monopole antenna would fit inside. For this reason there is a necessity to bend the monopole antenna and therewith to work with the Inverted L Antenna (ILA) - Fig. 5.20b.

5.3.3 b Analysis and Improvement of the WU Performance

The given configuration with both short and long ILA did not satisfy the requirements from Chapter 5.3.2. These Wheel Units have either problems with the level of the radiated power or with performance stability over the required temperature range. In such a case there was a necessity of technical advancement of the present WUs.

The analysis of the contemporary WUs and performance improvement followed the underneath listed stages:

- Measurement analysis of the present WUs.
- Simulation analysis of the present WUs - electromagnetic model.
- Employed embedding and housing material characterization - determination of electric material properties.
- Improvement and revision of the present WU RF performance.
- New WU proposal.

The already stated physical difference between both present WUs was the antenna length. Both units utilize the same module with the same embedding material, the same housing and PCB as well as battery architecture. Second difference was the matching network concept. The WU with the long ILA features a two stage matching network (Fig. 5.21a).

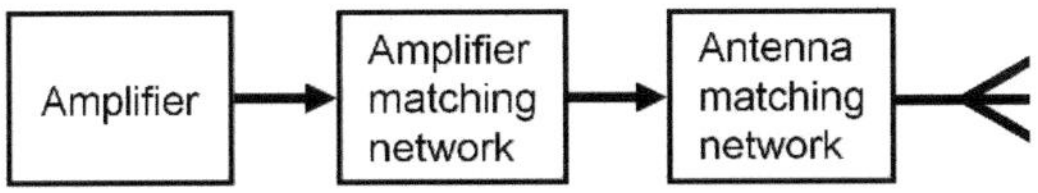

(a) Two stage matching network of WU with long ILA

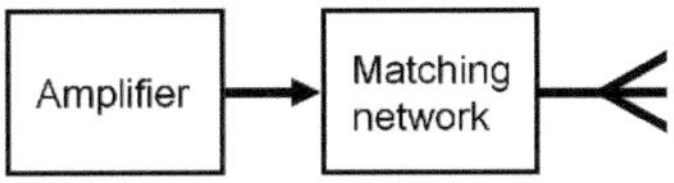

(b) Single stage matching network of WU with short ILA

Figure 5.21: Matching network architecture of the existing Wheel Units

In this case there is a separate matching network for antenna and a separate one for the amplifier. Both amplifier and antenna are matched to 50 Ω. On the contrary WU with short ILA features a one stage matching network (Fig. 5.21b): matching of the antenna directly to the amplifier impedance.

The matching networks corresponding to the above described architectures are documented with all lumped elements in Fig. 5.22. Fig. 5.22a for WU with long ILA, Fig. 5.22b for WU with short ILA.

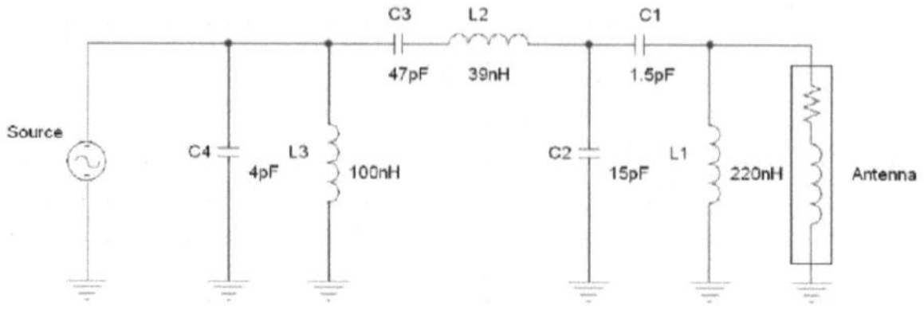

(a) Matching network of the WU with long ILA

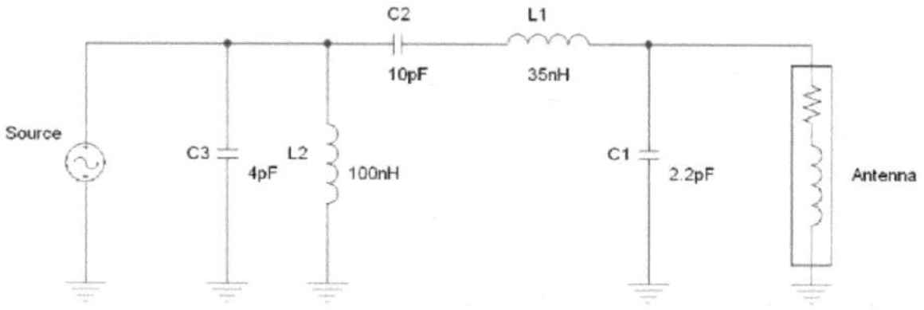

(b) Matching network of the WU with short ILA

Figure 5.22: Matching network of the existing Wheel Units

The two stage matching network comprises elements $L1$, $C1$ and $C2$ used to match the antenna impedance to 50 Ω and elements $L2$, $C3$, $L3$ and $C4$ used to match the source impedance to 50 Ω. Single stage matching network utilizes elements $C1$, $L1$ and $C2$ to match the antenna directly to the amplifier impedance.

After documenting the physical design differences, the first examination to follow was to measure the actual performance of both WUs (Table 5.3). Both radiated power and loss in the radiated power over temperature range are measured.

Parameter	Temperature	Requirement	WU with long ILA	WU with short ILA
Radiated power	25 °C	- 12 dBm	- 15.3 dBm	- 17.2 dBm
Maximum acceptable loss over the temperature range	25 °C - 105 °C	< 3 dB	- 8 dB	- 2.3 dB

Table 5.3: Measured radiated power of the present Wheel Units

Interpreting the above results, WU with long ILA performs better than WU with short ILA at room temperature (25 °C). Considering the loss in radiated power over temperature range, WU with short ILA fulfils the requirements with loss less than 3 dB, whereas WU with long ILA looses 8 dB, which exceeds the requirements.

Further, the geometry analysis of the present Wheel Units was conducted in order to verify and find the reasons for different performance over the given temperature range of the Wheel Unit equipped with short (104 mm) and long (146 mm) ILA. The electromagnetic simulation model of the WU was prepared. FEKO simulation models of the original 104 mm (short ILA) and 146 mm (long ILA) antenna are depicted in Fig. 5.23. In the presented model embedding material and housing are not visible.

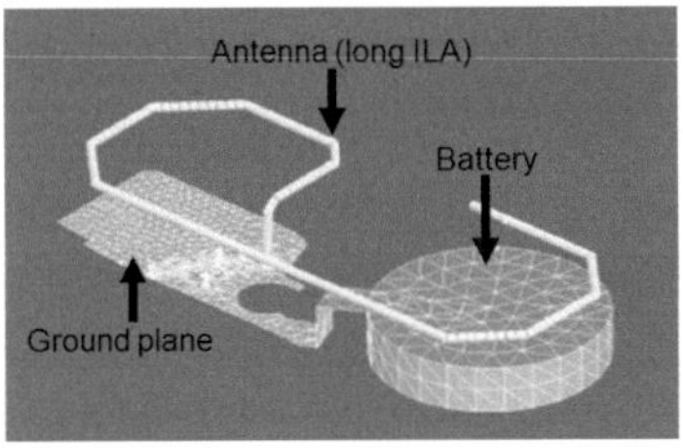

(a) WU with long ILA antenna (146 mm)

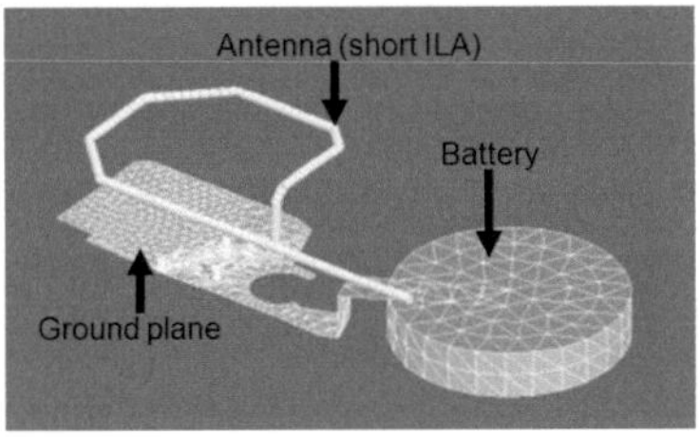

(b) WU with short ILA antenna (104 mm)

Figure 5.23: WU simulation models with monopole antenna (FEKO)

The above presented simulation models were used in order to examine the antenna performance in respect to the antenna length. Also battery influence was taken into account: battery - less and a model with battery presence was simulated. Results are gathered in Fig. 5.24.

It was proved that for battery - less antenna model, the longer the antenna, the higher efficiency and radiation resistance (Fig. 5.24a, Fig. 5.24c) and the lower the loss resistance are (Fig. 5.24b). The situation is however different with the considered battery presence in the model. With the longer antenna, efficiency and radiation resistance decreases whereas loss resistance increases.

The battery influence may be also additionally demonstrated with the observation of the electric field distribution within the WU module (Fig. 5.25). There is strong capacitive coupling between the long antenna and the battery. The magnitude of the electric field (E) between the battery (which is connected directly to the ground plane) and the antenna is higher for the long ILA (Fig. 5.25a) compared to the short ILA (Fig. 5.25b).

From the above, it may be easily deducted that the length of 146 mm is not optimal for the employed ILA antenna. From Fig. 5.24 it is to be read that for shorter lengths (< 120 mm), the antenna efficiency, as well as the radiation resistance increases, whereas loss resistance drops in the battery presence.

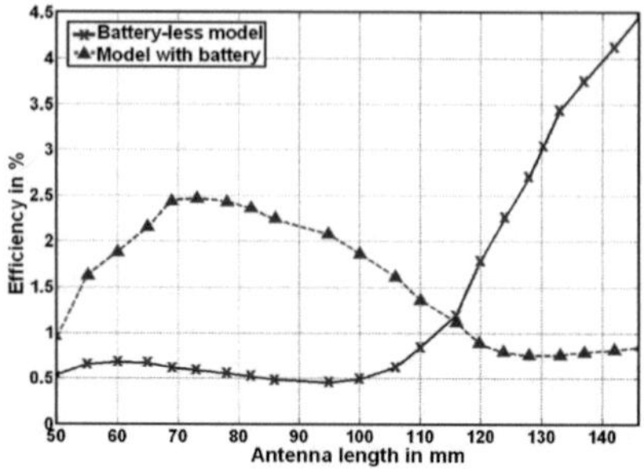

(a) Efficiency versus antenna length

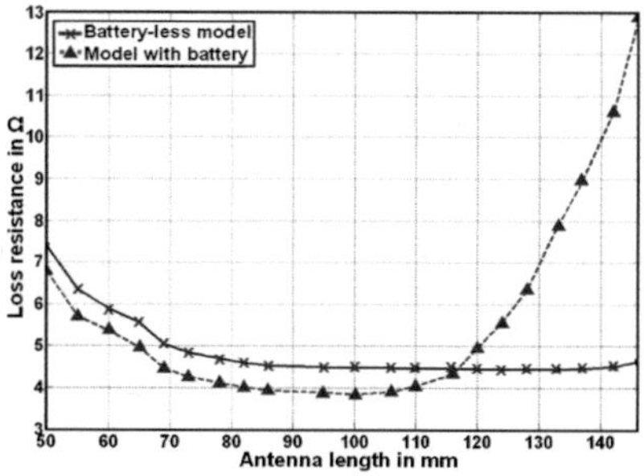

(b) Loss resistance versus antenna length

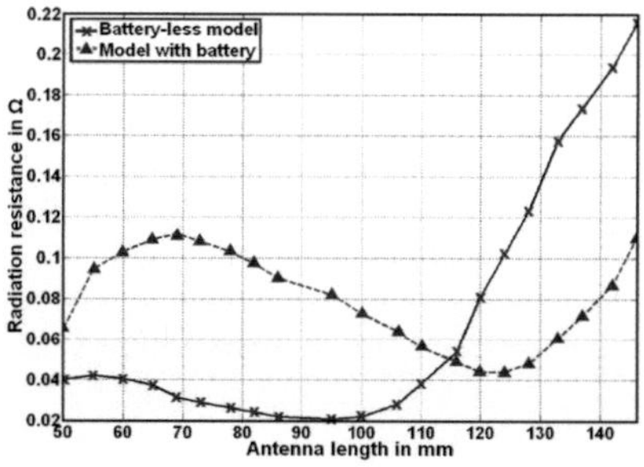

(c) Radiation resistance versus antenna length

Figure 5.24: Influence of the battery on the Wheel Unit monopole antenna performance at 433.92 MHz

Further, effect of the temperature change on the antenna gain for both WUs was considered. This was analysed with the means of electromagnetic simulation and according to Fig. 4.25 the electric material properties (permittivity and loss factor) were adjusted to the values typical for each temperature range.

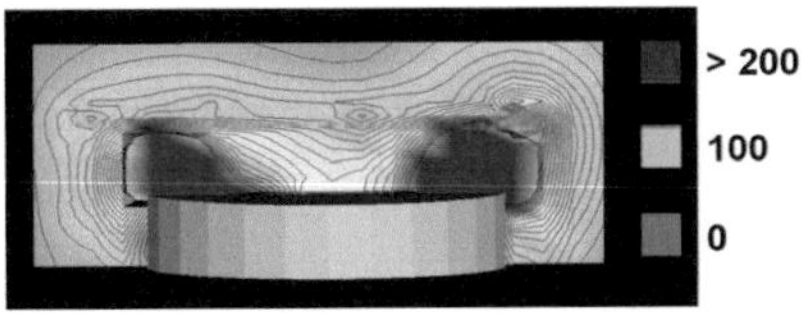

(a) Wheel Unit with long ILA

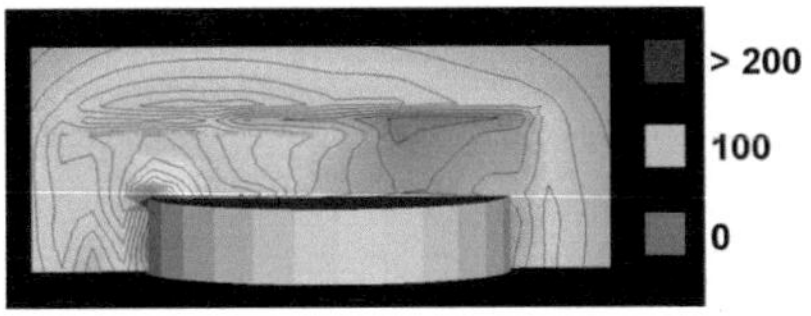

(b) Wheel Unit with short ILA

Figure 5.25: Simulated electric field distribution E_t in V / m, cross section with the battery

The simulated average antenna gain is depicted in Fig. 5.26. The frequency of 433.92 MHz is tagged and labelled with a vertical arrow. For both antennas, the gain loss for higher temperatures amounts up to 4 dB.

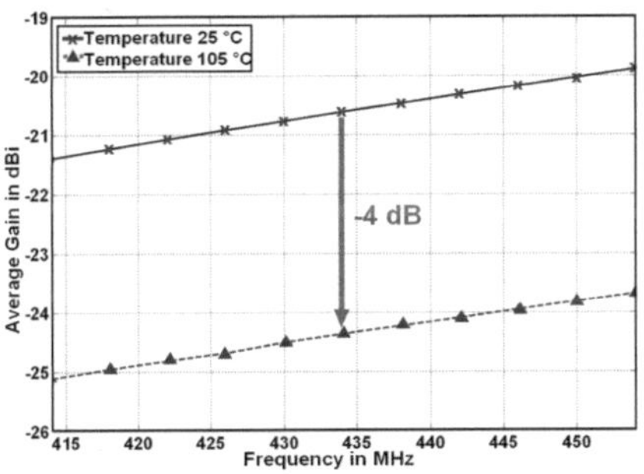

(a) WU with long ILA

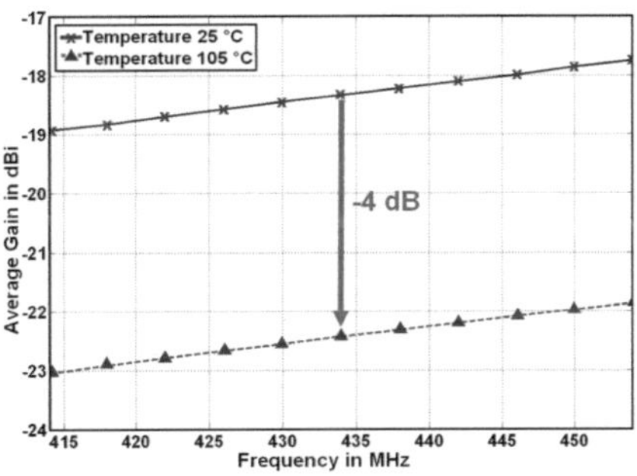

(b) WU with short ILA

Figure 5.26: Temperature effect on the average ILA gain - simulation

This loss results primarily from the temperature dependent change in the housing and embedding material. Material properties (loss factor increment) and therefore the effect is identical for both antennas. Specified maximum acceptable loss over temperature range should not exceed 3 dB. (Table 5.2)

Next step was the analysis of the already presented matching networks for both long and short ILA (Fig. 5.22). In order to state whether the matching introduces any losses and to determine how high these losses could be, the transmission coefficient was simulated both for 25 °C and 105 °C (Fig. 5.27). For the simulation purposes a high value of the quality factor for the lumped components was considered and kept constant for both temperature ranges.

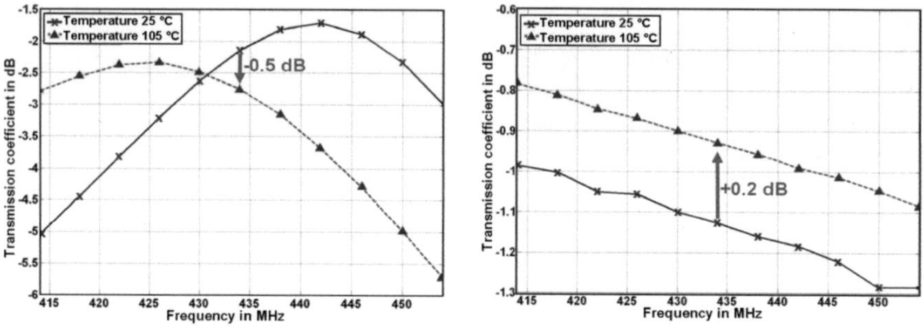

(a) WU with long ILA

(b) WU with short ILA

Figure 5.27: Matching network transmission coefficient - simulation

For the long ILA the matching network causes a loss of 2.3 dB at 25 °C and 2.8 dB at 105 °C. This means increased temperature causes additional loss of 0.5 dB.

For the short ILA with direct matching network, the losses are lower: 1.12 dB at 25 °C and 0.93 dB at 105 °C. This means that in the case with increased temperature, the losses are reduced by approximately 0.2 dB.

After having analyzed the antenna element influence and having the knowledge about the matching network behaviour for both temperature ranges, it is evident to analyze the RF transmitter output, which is influenced by an internal amplifier. The block diagram of the PLL transmitter from ATMEL is shown in Fig. 5.28 [71].

The ATMEL transmitter ATA5756/ATA5757 has been developed especially for demands of RF transmission systems like TPMS. The last output stage of the implemented transmitter is the power amplifier. The output power depends on the impedance seen by the amplifier. In order to simulate the output power a PSPICE model was prepared [72]. The result of this simulation is in the 3D plot projected on the smith chart in Fig. 5.29.

According to the simulated values the maximum output power amounts up to 7 dBm. For better visualization a 2D contour plot is depicted in Fig. 5.30. Each contour represents change of 1 dB. The impedance where the transmitter delivers maximum power is marked with a cross, this is for impedance value of (280 + j310) Ω.

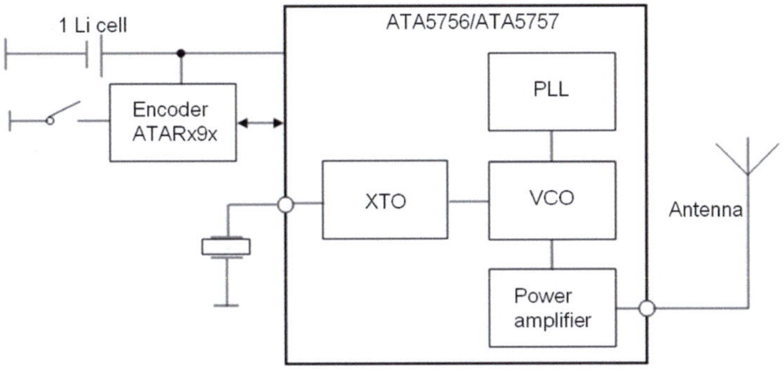

Figure 5.28: Block diagram of PLL transmitter IC ATA5756/ATA5757

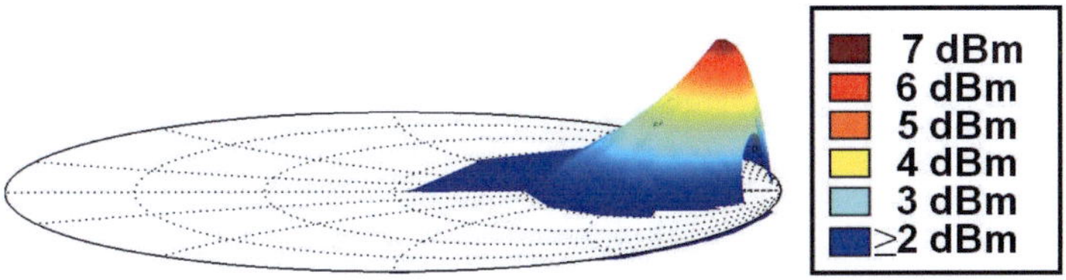

Figure 5.29: Simulated output power of PLL transmitter IC ATA5756/ATA5757

Further the amplifier output power was plotted on the Smith Chart (Fig. 5.31). Impedance was measured and simulated for the WUs with both long and short ILA. A very good correlation between simulated and measured values may be observed. The frequency range between 414 MHz and 454 MHz was taken into account, frequency of interest of 432.92 MHz is marked in each case with arrows.

From Fig. 5.31a it can be read that at 25 °C the antenna is fed with 6 dBm and at 105 °C with 4.5 dBm. For WU with short ILA (Fig. 5.31b) the antenna receives 0 dBm at 25 °C and 4 dBm at 105 °C.

Summing up, the antenna in the WU with long ILA receives more output power from the amplifier but with higher temperature 1.5 dB are lost. Whereas for the WU with short ILA, with higher temperature, the amplifier output power rises.

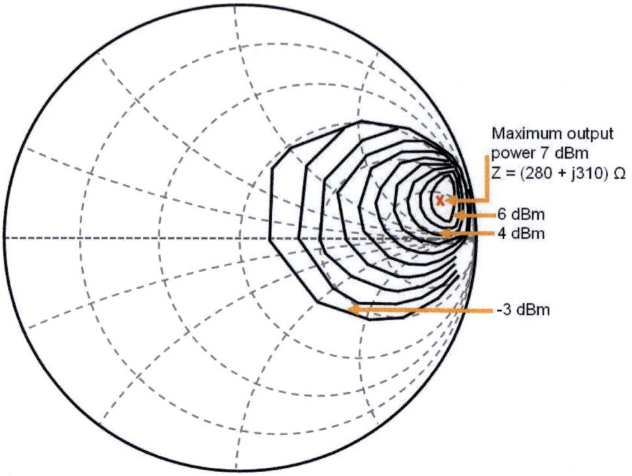

Figure 5.30: Simulated output power of PLL transmitter IC ATA5756/ATA5757 on Smith Chart

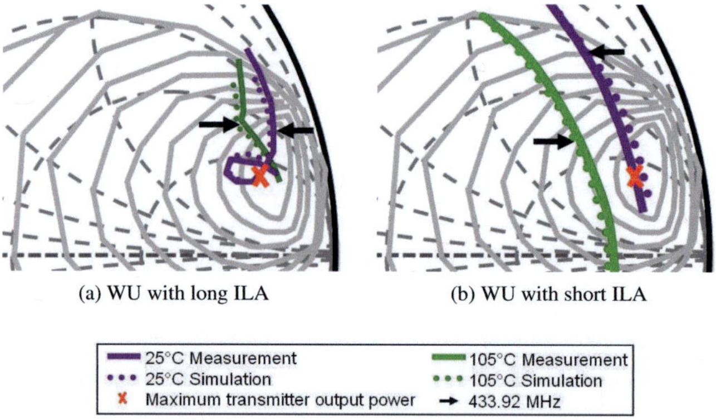

(a) WU with long ILA

(b) WU with short ILA

Figure 5.31: Simulated output power of the transmitter. Measured and simulated load impedance

After identifying the values for the delivered power from the transmitter to the antenna element together with the matching network for both WUs equipped with long and short ILA, the last comparison is the calculation of the radiated power.

The radiated power is calculated according to Eq. 5.5.

$$P_{rad} = Average\ antenna\ gain - Matching\ network\ loss + Amplifier\ output\ power \quad (5.5)$$

The simulated results are gathered in Table 5.4 for WU with long ILA. Table 5.5 comprises simulated results for WU with short ILA.

Parameter	Temperature 25°C	Temperature 105°C	Difference due to temperature change
Antenna gain	-20.5 dBi	-24.5 dBi	4 dB
Loss from matching	-2.3 dB	-2.8 dB	0.5 dB
Output power of amplifier	6 dBm	4.5 dBm	1.5 dB
Radiated power	-16.8 dBm	-22.8 dBm	6 dB

Table 5.4: Simulated RF parameters for WU with long ILA

Parameter	Temperature 25°C	Temperature 105°C	Difference due to temperature change
Antenna gain	-18.5 dBi	-22.5 dBi	4 dB
Loss from matching	-1.12 dB	-0.93 dB	0.19 dB
Output power of amplifier	0 dBm	4 dBm	4 dB
Radiated power	-19.6 dBm	-19.4 dBm	0.2 dB

Table 5.5: Simulated RF parameters for WU with short ILA

In further step the obtained simulation and calculation results are compared with measured values. Radiated power for both WUs was measured with the GTEM cell. GTEM cells find application for the measurement of the average radiated power of small antennas [73]. In Table 5.6 simulated and measured RF parameter values are compared. Simulation accuracy is very high, what confirms the precision of the prepared antenna models.

Parameter	Temperature	WU with long ILA		WU with short ILA	
		Meas.	Sim.	Meas.	Sim.
Radiated power	25°C	-15.3 dBm	-16.8 dBm	-17.2 dBm	-19.5 dBm
	105°C	-23.3 dBm	-22.8 dBm	-19.5 dBm	-19.5 dBm
Loss in radiated power	25°C to 105°C	-8 dB	-6 dB	-2.3 dB	0 dB

Table 5.6: Comparison of simulated and measured RF parameters for WU with long and short ILA

For both temperature ranges the gain of the antenna in WU with short ILA is higher than the gain of the antenna in the WU with long ILA. Shortening the antenna arm, showed advantageous effect on RF antenna performance as in the case of short ILA the capacitive interaction with the battery is reduced.

Performance of the WU with short ILA shows more stability in the whole temperature range. In WU with short ILA, the output power of the transmitter amplifier for temperature of 105 °C increases by 4 dB.

None of the WUs satisfies the requirement of the radiated power level of - 12 dBm at 25°C and minimum radiated power of -15 dBm at 105 °C (Table 5.2). Losses due to temperature change are much lower for short ILA with 2.3 dB than for long ILA with 8 dB.

5.3.4 Inverted F Antenna Development

After the analysis of the present WUs, the further task was to advance the antenna performance in terms of:

- Improvement of the radiated power by 5 dB - antenna topology and WU components. Due to battery life time the consumption power is to be kept constant.
- Reduction of dependency of antenna performance on temperature change.

Summarizing following optimization steps were analysed and subsequent changes concerning following WU elements were introduced:

- New antenna concept including change of antenna material.
- Change of the embedding material.
- Matching network influence analysis.

The structure and main advantages of the new antenna type, Inverted F Antenna (IFA), are going to be shortly presented as this antenna type was implemented in the modified Wheel Unit.

Inverted F Antenna gained on popularity in mobile communication systems [75] and automotive applications due to its properties. This antenna type is particularly known for its capabilities to allow simple impedance match still within a low profile design. Second characteristic is the ability to originate both vertically and horizontally polarized electric fields [76].

5.3.4 a Fundamental Structure of Inverted F Antenna

IFA is an enhancement of the already presented ILA. Similarly, as ILA resembles an inverted L letter, IFA resembles an inverted F letter and therefore the name of the antenna.

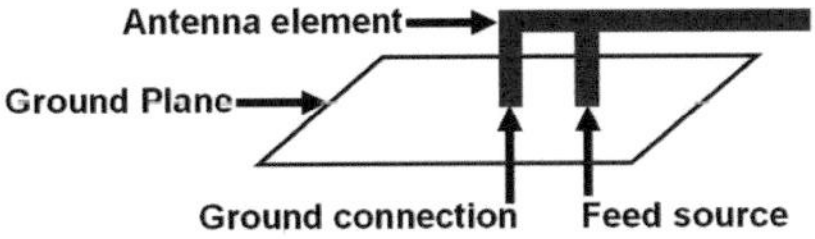

Figure 5.32: Structure of the Inverted F Antenna

IFA is effectively a quarter wavelength ($\lambda/4$) monopole in which (as in ILA) a portion of the antenna length has been bend so it is parallel to the ground plane. IFA height is usually of approximately one tenth of a wavelength ($\lambda/10$). Therewith it is about 2.5 times lower than the (perpendicular to the ground plane) quarter wavelength monopole [80]. Additionally at IFA a short circuit (inductive) stab is introduced. It compensates the increased capacitance that originates from the parallel antenna part to the ground [77]. Fundamental IFA structure is depicted in Fig. 5.32

5.3.4 b Modification of the Present Wheel Unit

The implemented IFA structure (Fig. 5.33) differs from its fundamental form. In order to omit the battery influence (Fig. 5.25) the antenna wire was folded.

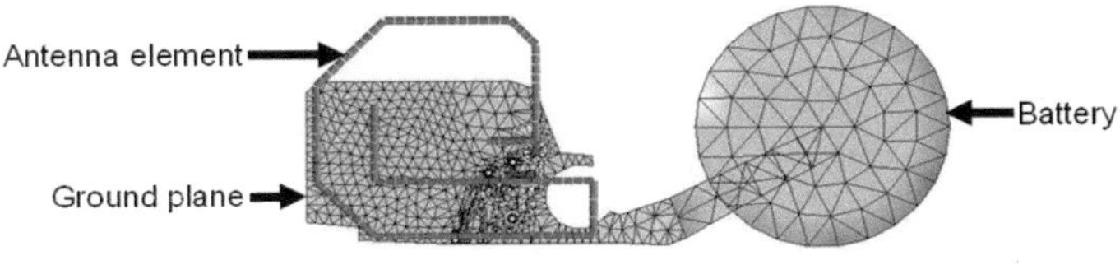

(a) Top view

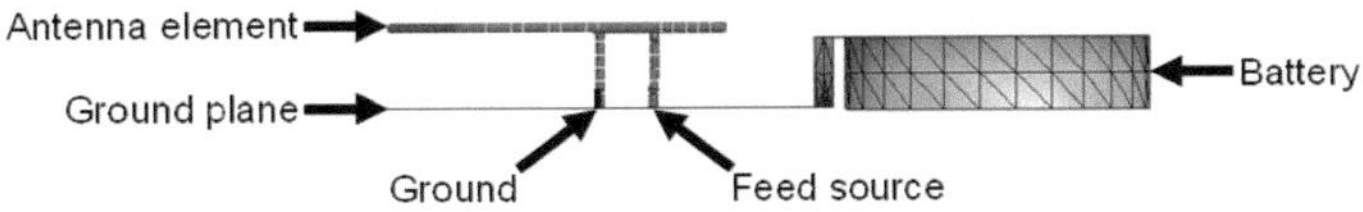

(b) Side view

Figure 5.33: Simulation model (FEKO) of realized IFA antenna

Depicted FEKO simulation model comprises antenna element, ground plane and battery, embedding material and housing are hidden.

The whole WU assembly is illustrated in Fig. 5.34. Here also embedding material as well as housing are visible. The overall WU size is given in respect to the wavelength. The simulation model was originated in HFSS.

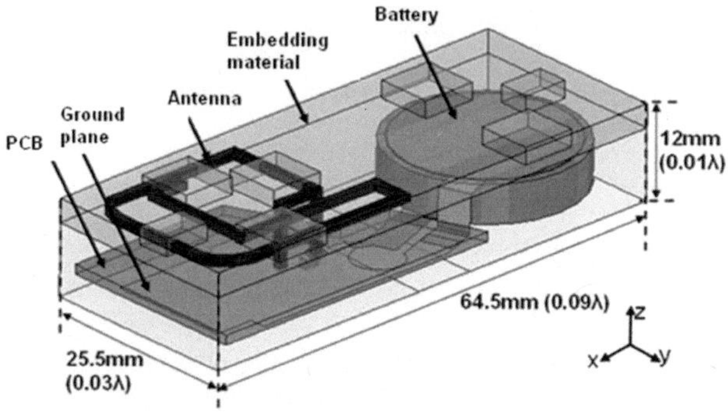

Figure 5.34: Simulation model (HFSS) of the realized WU with IFA

Apart from the antenna structure modification, also the antenna material was changed. Due to better conductivity stainless steel (ρ = 2.8 exp10 S/m) was exchanged by copper (ρ = 5.7 exp10 S/m).

Further, a third optimization step was the replacement of the embedding material. The originally used polyurethane is a lossy material with unstable behaviour when temperature change occurs (tan δ = 0.03 for 25 °C and tan δ = 0.08 for 105 °C, Chapter 4.8, Fig 4.25). Material of interest should preferably reveal stable characteristic for the demanded temperature range as well very low, negligible losses. Moreover the material should be permitted for automotive use.

Due to above listed demands chosen material was silicon. Measured electrical material properties of silicon in comparison to polyurethane are presented in Chapter 4.8, Fig. 4.26. For both extreme temperatures out of required temperature range, silicon material losses are negligible (tan δ = 0). the WU with polyurethane and silicon was photographed and are presented in Fig. 5.35. The only optical difference is the colour of both materials.

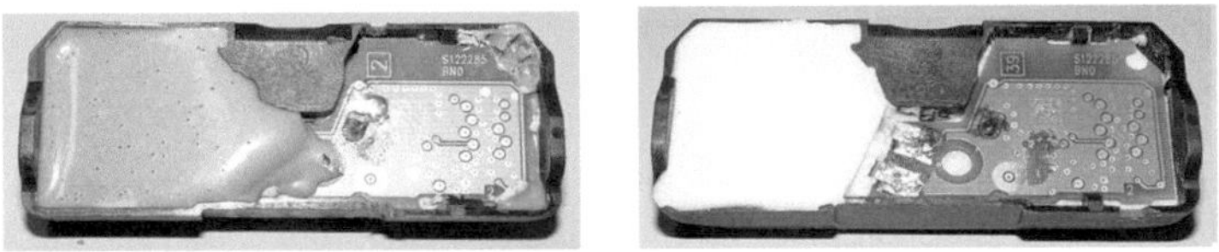

(a) WU with polyurethane (b) WU with silicon

Figure 5.35: Change in the embedding material from polyurethane to silicon

5.3.4 c Analysis and Improvement of the Modified WU Performance

As electromagnetic model of the WU with new IFA antenna was already presented, it is now to present the achieved results and check whether this WU brings better RF performance.

Simulated and measured input impedance of WU with IFA is documented in Fig. 5.37a. In order to match the antenna, a two stage (Fig. 5.21a) matching network was designed - schematics of the matching network is shown in Fig. 5.36.

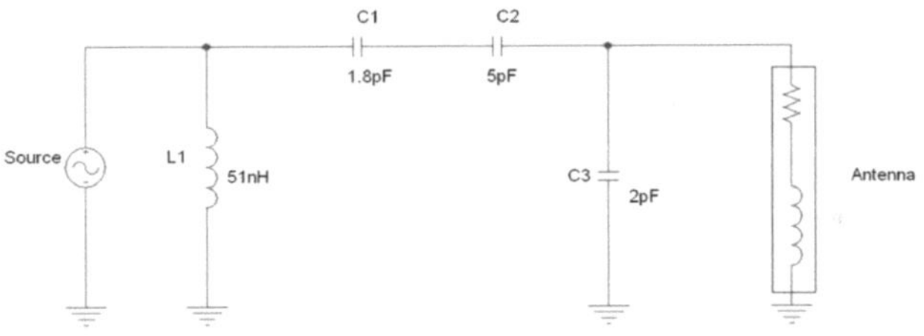

Figure 5.36: Matching network of the WU with IFA

Components $L1$ and $C1$ were used to transform the output impedance of the amplifier to 50 Ω and components $C2$ and $C3$ were used to match these 50 Ω to the antenna. Achieved reflection coefficient was firstly simulated and then measured (Fig. 5.37b) with very good result agreement.

It is very convenient to work with antenna and amplifier both matched to 50 Ω but it was very important to state that the matching network does not introduce any additional loses. WU efficiency was simulated for two cases: antenna element both with and without matching network (Fig. 5.38).

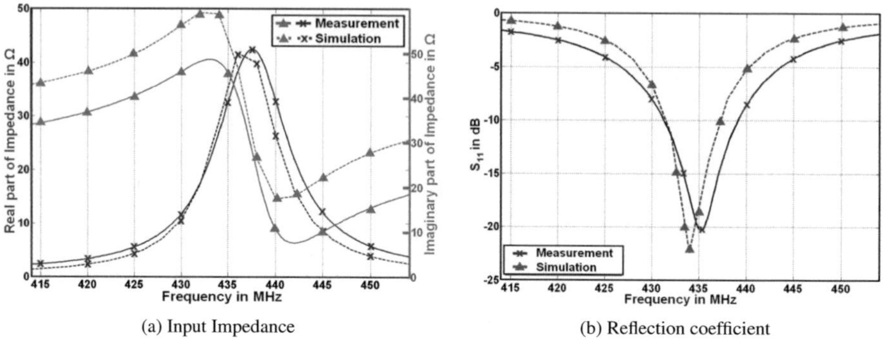

Figure 5.37: Simulated and measured impedance and input reflection coefficient of the WU with IFA

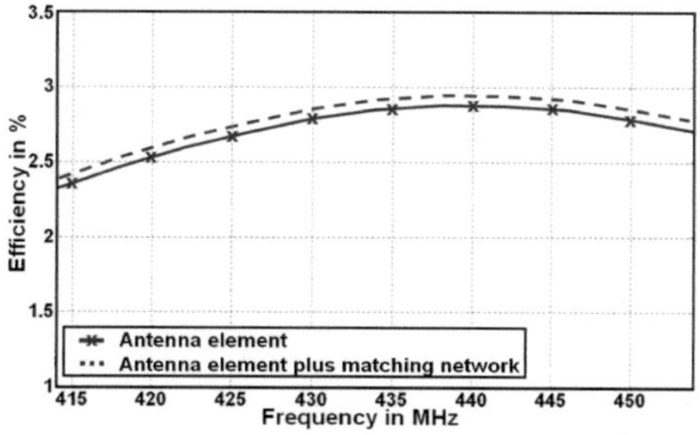

Figure 5.38: Simulated efficiency of the WU with IFA

According to the above results the effect of the two stage matching network seems to be negligible. It was important also to check whether this is the case for both temperature ranges. In Fig. 5.39, the simulated matching network transmission coefficient is depicted for 25 °C and 105 °C. The foreseen matching network causes slight, negligible loss at room temperature as well as at high temperature.

The next step was to check the amplifier output for both temperature ranges. As already mentioned the output power of the amplifier varies depending on the load impedance connected to it. Further the amplifier output power was plotted on the smith chart together with the load impedance seen by it, what is illustrated in Fig. 5.40a.

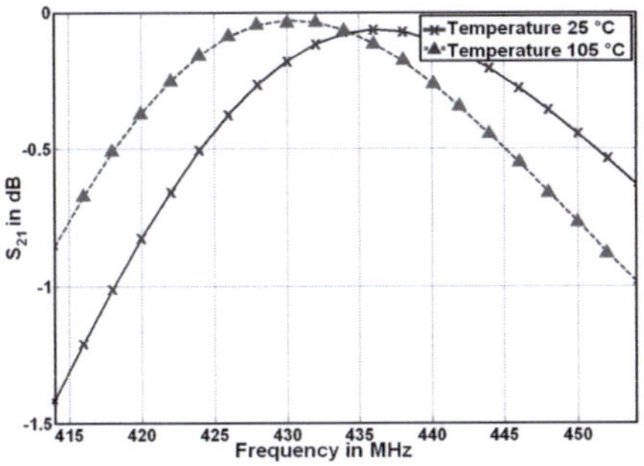

Figure 5.39: Simulated matching network transmission coefficient of the WU with IFA

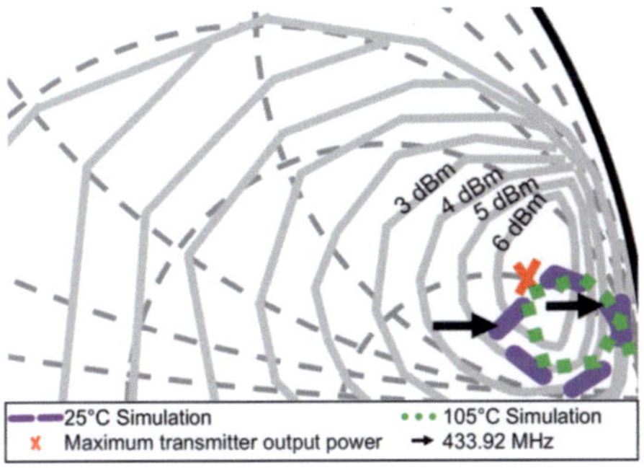

(a) Simulated output power of the transmitter and simulated load impedance for WU with IFA

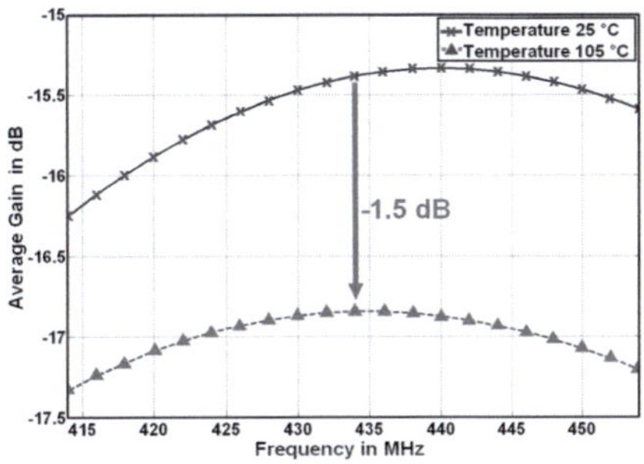

(b) Simulated average gain of the WU with IFA

Figure 5.40: Simulated temperature dependance of the characteristic parameters of the WU with IFA

It can be seen that the amplifier transfers appr. 5.5 dBm power to the antenna at 25 °C and appr. 4 dBm at 105 °C. That means, over the temperature the amplifier contributes 1.5 dB loss in the radiated power of the Wheel Unit.

Antenna average gain was compared also at room temperature with maximum temperature of 105 °C. As in Fig. 5.40b it may be observed that the loss in gain is 1.5 dB which expresses exactly the difference in the amplifier output power for different temperature ranges. This confirms negligible losses from the matching network.

Parameter	Temperature 25°C	Temperature 105°C	Difference due to temperature change
Antenna gain	-15.3 dBi	-16.8 dBi	1.5 dB
Loss from matching	0.2 dB	0.1 dB	0.1 dB
Output power of amplifier	4.3 dBm	4 dBm	1.5 dB
Radiated power	-9.6 dBm	-12.6 dBm	3 dB

Table 5.7: Simulated and calculated RF parameters for WU with IFA

5.3.4 d Radiation Pattern of Wheel Unit with Inverted F Antenna

As final WU test the radiation pattern was measured in three planes and results are presented in Fig. 5.41.

Such a measurement not only delivers information on the radiation pattern but also approves the accuracy of the created simulation model. To fulfill the accuracy constraint, the simulation and measurement results must be aligned.

Characteristic measured radiation pattern values for the Wheel Unit with Inverted F Antenna are listed in Annex B, Table B.7, which confirms accuracy of the simulation model at level of 1 dB. Therewith and upon above presented results it may be stated that a very good correlation between measured and simulated values was achieved.

5.3.5 RF Performance Comparison of WUs with Monopoles and Inverted F Antenna

The performance of the WUs equipped with long ILA, short ILA and IFA were analyzed. For final conclusions only the radiated power by the WUs is compared. The results together with the requirements are listed in the underneath Table 5.8 It may be easily seen that the

Condition	Radiated power [dBm]				Radiated power improvement in dB
	Specifications	WU with long ILA	WU with short ILA	WU with IFA	
25 °C	>-12	-15.5	-17.2	-9.6	5.9
105 °C	>-15	-23.5	-19.5	-12.6	10.9

Table 5.8: Performance comparison of Wheel Units under investigation (long ILA, short ILA, IFA) at room and high temperature

WU with the long ILA, which was already present did not satisfy the required radiated power levels and also lost more than 3 dB due to temperature change. The new WU satisfies the RF requirements and therefore it may be further implemented in TPMS system.

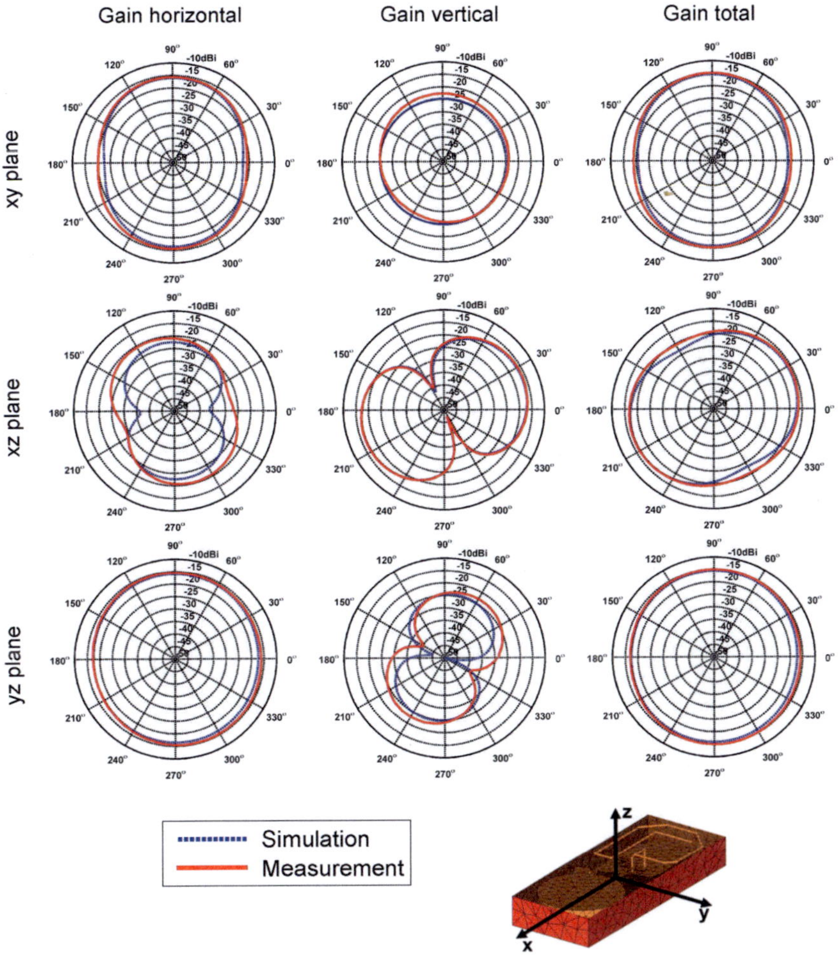

Figure 5.41: Measured and simulated radiation pattern of WU with IFA in free space

5.3.6 Loop Antenna Development

Chapter 5.3.6 presents a new WU design. The overall WU concept, was kept as valve mounted module. The newly designed WU features smaller size and a new radiating element form. The RF analysys follows.

The basic representation of the loop antenna is depicted in Fig. 5.42. Generally loop may be of any shape: circular, rectangular, triangular, rhombic or square. It is a radiating coil on a ferrite or air core consisting of one (Fig. 5.42a) or more turns (Fig. 5.42b).

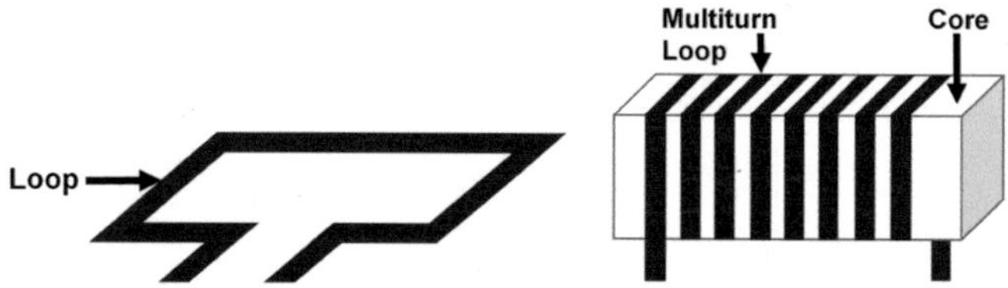

(a) Single loop (b) Multiturn loop

Figure 5.42: Fundamental structure of the loop antenna

Classified on the size basis, there are two types of loop antennas: in the first type the physical dimensions of the loop are comparable with one wavelength and in the second type the dimensions of the loop are very small compared to the wavelength [79]. For this work only small loop antennas are of interest.

For loops of small size, the radiation characteristic is independent on the loop inherent shape. Under such condition the radiation pattern of the loop is similar to that of an elementary dipole. Fundamental difference is the interchange of electric and magnetic field vectors. Therefore small loop antennas are often called magnetic dipole.

5.3.6 a Assembly of the WU with Loop Antenna

For the future systems a new alternative to the already existing Wheel Units was under research. The goal was to reduce the WU size and to improve the RF performance, particularly when interacting with rim and tyre. Having already analysed very carefully the basic antenna types such as the implementation of a monopole antenna and an IFA, a new antenna type was applied. The new WU works with a loop antenna.

The new WU is presented in Fig. 5.43a and belonging new antenna element in Fig. 5.43b. The fundamental WU assembly, with the same major component functionalities, is very similar to the previously discussed WUs (ILA, IFA).

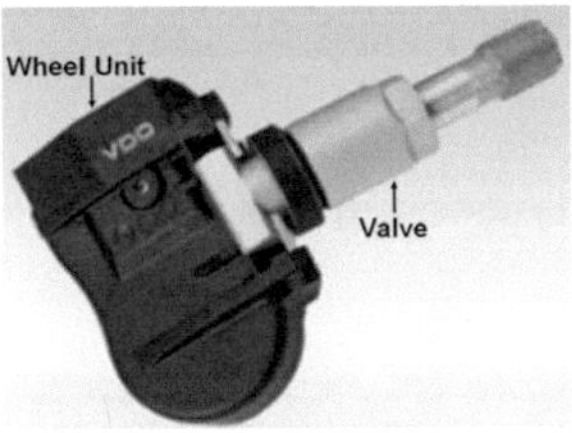

(a) WU with valve

(b) Antenna element

Figure 5.43: New Wheel Unit major components

Merely component mechanical- and RF design was slightly changed:

- There is a new WU housing, which was reduced to the size of 48 mm x 27 mm x 12 mm. Therewith the overall WU dimensions are reduced by 16 mm in the length and 2 mm in the width. The WU height, with 12 mm, is the same as in previous design of WU with the monopole or IFA antennas. In relation to the frequency of operation, the WU dimensions represent now respectively 0.07 x 0.04 x 0.02 of the wavelength.

  The main advantage, from an industrialization point of view, associated with the volume reduction is the resulting component weight decrease. Chosen housing material was PA66. At room temperature the electric permittivity of this material at 433.92 MHz is $\epsilon_r = 3.4$ and loss factor is $\tan\delta = 0.0076$ (Fig. 4.25).
- New PCB was designed with slightly smaller dimensions determined by smaller housing size. New PCB size is 20 mm x 24.2 mm. Carrier material is standard multilayer FR4 material with four electrical layers (Fig. 5.45).
- The battery is the same as in previous WUs (Fig. 5.19c).
- The antenna element (Fig. 5.43b) has got a form of a loop, realized with irregular pentagon. The loop biggest dimension is along the PCB - 21.6 mm. The height is limited by the WU housing - it is 7.35 mm. The loop circumference amounts λ / 20, therewith it fulfills the criterion and belongs to small antennas (Chapter 5.1).

  Mechanical benefit for this antenna solution, apart from stability, is easier producibility due to the fact that more accurate antenna positioning in respect to the PCB and housing is possible in the production process.

5.3.6 b Simulation Model of the WU with Loop Antenna

After introducing all the basic antenna elements, considerations have to be made how to model the WU, how to simplify the detailed design and to which extent. Two WU models were prepared:

- Very accurate model, where all PCB layers were taken into account (Fig. 5.44a).
- Simplified model with one metallic sheet representing the PCB (Fig. 5.44b).

Both models are depicted without housing and embedding material.

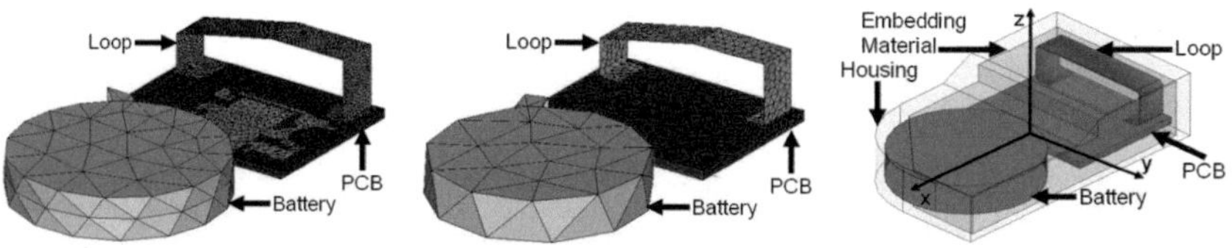

(a) Detailed FEKO model (b) Simplified FEKO model (c) Simplified HFSS model

Figure 5.44: Simulation model of the realized WU with loop antenna

Detailed and simplified FEKO models show only the radiating element and the PCB. The first of these models features a complete PCB structure with four layers (the complexity of this model reveals Fig. 5.45 which presents the layer structure). In order to present the whole structure of the WU with housing and embedding material, all elements are visible in HFSS simulation model in Fig. 5.44c.

It is generally known that the more detail, the more elements one model features, the longer the resulting simulation time. If possible it is naturally advantageous to work with very simplified models. It has to guaranteed however that these deliver correct results. This is the point for beneath considerations.

5.3.6 c Model Accuracy of the WU with Loop Antenna

For further examinations of the WU with loop antenna on the rim, with tyre and whole vehicle it would be advantageous or even necessary to use the simplified model. The task is to check whether this is possible.

The simulation time for detailed and simplified FEKO model was compared. It appeared, that the difference in the simulation time for both models is very meaningful. Compared are simulation times on a cluster PC with 10 parallel CPUs for each simulation. Prior to this comparison an analysis of the stability (convergence) of each of the models was carried out [37] - this is necessary and very important for models of very small size and featuring small elements (much smaller than λ / 10). The basis for the simulation time prediction and finally simulation execution is the overall number of the basis elements dielectric and metallic triangles. Simplified model was simulated within 5 minutes whereas an accurate model needed over one hour. Comparison of the data for both models is listed within the Table 5.9.

WU with loop antenna - model type	Simplified	Accurate
Number of metallic triangles	1850	12740
Number of dielectric triangles	3316	5940
Triangle total number	5174	18680
Number of basis functions for MoM	13800	39305
Simulation time	5 min	1h 25 min
Number of CPUs	10	10

Table 5.9: Comparison of the simulation time for WU with loop antenna for the simplified and the accurate model

It is clear, that further considerations of the full TPMS system would be feasible only with the simplified model that requires much less simulation time. Further step is to compare and check the accuracy of the characteristic output parameters of both models versus measured values.

5.3.6 d Radiation Pattern of Wheel Unit with Loop Antenna

Radiation pattern was measured and accordingly simulated in three planes. The orientation of the model may be read from Fig. 5.44c and is also indicated in Fig. 5.49 with the achieved results.

Comparison of the simulation results revealed very similar model performance in regard to 3D radiation pattern with the difference in the amplitude of the radiated power. In order to be able to use the simplified model, is has to be analyzed where the discrepancy in the radiated power comes from. For this purpose a detailed analysis of the matching network will follow.

5.3.6 e Simulation Model of the Wheel Unit with Loop Antenna

The discrepancy in the simulation results between simplified and detailed model is structure based (simplification of the model and matching network). The matching network of the loop antenna will be thoroughly analyzed. A stable matching network with very small losses is very important for the optimum performance of the small loop antenna and it is important to examine how it contributes to the radiated power.

In order to analyze the mutual influence of the input impedance measurement uncertainty and quality of the matching network with certain quality of its components, an additional network simulation model was prepared with the help of Ansoft Designer.

The investigation is divided into four following steps: introduction of matching network model (step 1), considerations on the measured and simulated real part of the impedance (step 2), circuit simulation model of the matching network (step 3), analysis of the achieved results (step 4). Explanation of the steps follows, where it is explained which values were used and how the analysis was done.

Firstly PCB and matching network model are introduced. All PCB layers with indiction on the matching network area are depicted in Fig. 5.45.

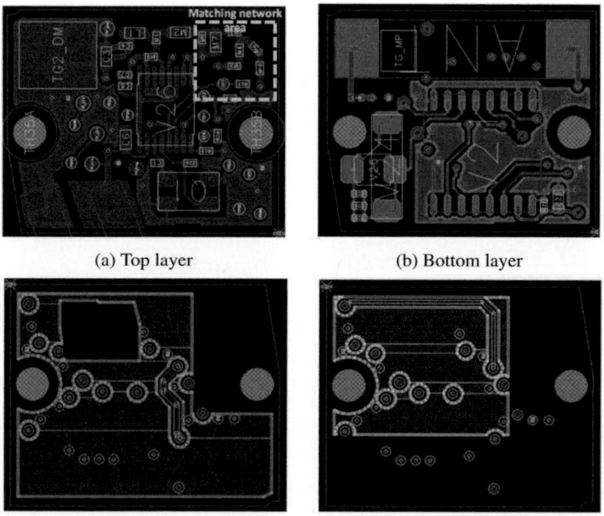

(a) Top layer (b) Bottom layer

(c) Inner layer 1 (d) Inner layer 2

Figure 5.45: PCB layers of the WU with loop antenna

Matching network elements as well as block diagram are presented in Fig. 5.46. The antenna feeding point is indicated by Port 2. Antenna input impedance is transformed to 50 Ω at Port 1. Matching network element values are to read from Fig. 5.46b.

As the second step the real part of the input impedance of the designed antenna has to be considered. It is very small and therefore very difficult for precise measurement. An analysis of the matching network behavior was done in respect to theoretical and measured antenna input impedance. In this way performance of the antenna and its matching network could be investigated.

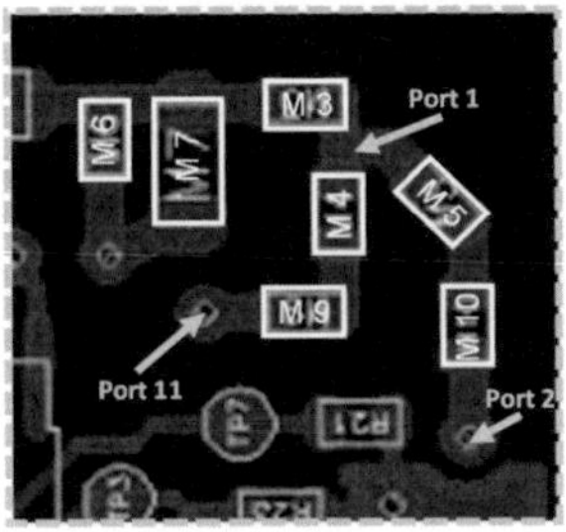

(a) PCB layout

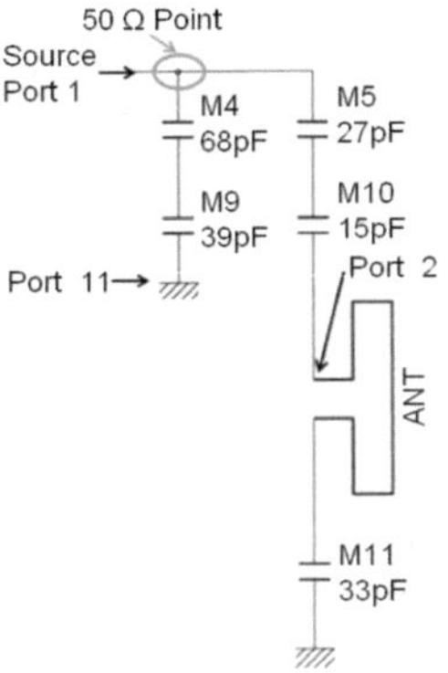

(b) Circuit diagramme

Figure 5.46: Loop antenna matching network

All values for the analysis are gathered in Table 5.10 and Table 5.11.

Antenna impedance	Theoretical / Simulated	Measured
Real part in Ω	0.5	1.1 to 1.4
Imaginary part in Ω	75	50

Table 5.10: comparison of the simulated and measured input impedance of the loop antenna

Simulation analysis	Assumed value	Assumed meas. uncertainty
Impedance real part in Ω	0.2 to 0.8	0.2 to 1
Element quality	According specification 300	Ideal 1000

Table 5.11: Listed values for consideration of the antenna and matching network interaction

Measured real part of the antenna impedance varied between 1 Ω to 1.4 Ω, whereas theoretical, simulated value in FEKO reached 0.5 Ω. Dependent on what could be a true real value of the input impedance, the interaction of the antenna with the designed matching network could vary and therewith contribute to either higher or lower loss. This interaction is a subject of the examination.

Table 5.11 gathers the theoretical possible supposed values. According to the above presented simulated and measured results, the value of the input impedance real part that was assumed for the analysis purposes was set from 0.2 Ω to 0.8 Ω and the according possible simulation and measurement error value 0.2 Ω to 1 Ω. Further also the quality of the matching network components is very important. According to the element specification, the quality factor value is 300 and theoretical, ideal value, if such elements would exist, would be 1000.

Third step is to introduce the circuit simulation model designed in Ansoft Designer where all previously discussed parameters may be controlled. The model is depicted in Fig. 5.47.

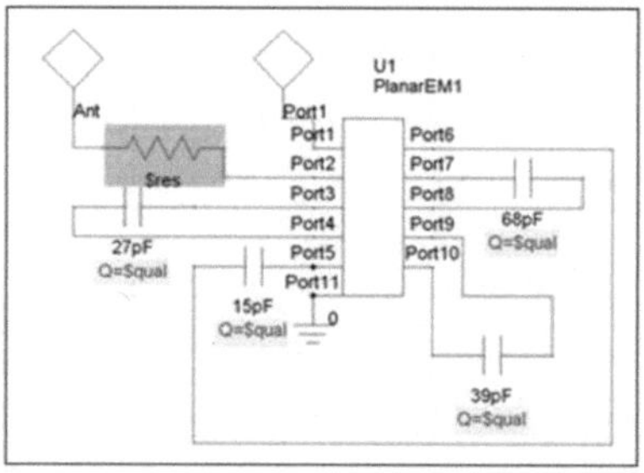

Figure 5.47: Matching network simulation model of the loop antenna

The introduced model has to be explained. Matching network was realized with the help of an eleven port element. This element represents the electric behavior of the PCB tracks connecting all the involved lumped elements of the matching network. Simulation model layout (Fig. 5.47) corresponds to the previously introduced loop antenna matching network from Fig. 5.46. Parameter Q ($qual) stands for the element quality values and is variated in the simulation process. Assumed antenna impedance values from Table 5.11 will be used. The according measurement uncertainty is introduced by parameter $res. The sum of the assumed antenna impedance and measurement error. Both parameters together are meant to represent the real antenna impedance in port 2.

The simulation was conducted with the above described parameters, were transmission and reflection factor was noted in frequency range from 300 MHz to 500 MHz. The results are presented in Fig. 5.48.

Last, fourth step is the interpretation of the above results and the implementation of the correction factor as well as decision about its value. Further 3D EM simulation processes are carried out in order to save simulation resources (time). Deviations from simple WU model simulation results can be corrected by the introduction of the correction factor. Simulated reflection factor varies between - 3 dB and - 14 dB and the transmission factor between - 3 dB and - 9.5 dB. Upon these values, the correction factor will be set at the level of 7.5 dB. Considering this factor, the simulation results, for the simplified model, of the radiated power would correlate with the measured values. Therewith the simplified WU loop model may be used for further investigations.

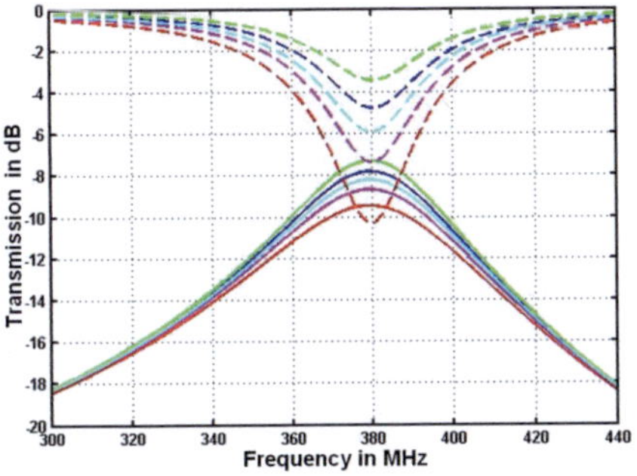

(a) ReZ = 0.2 Ω, ImZ = + j 50 Ω

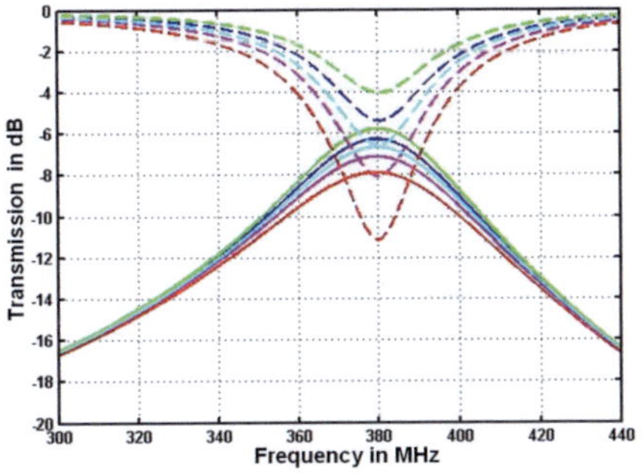

(b) ReZ = 0.3 Ω, ImZ = + j 50 Ω

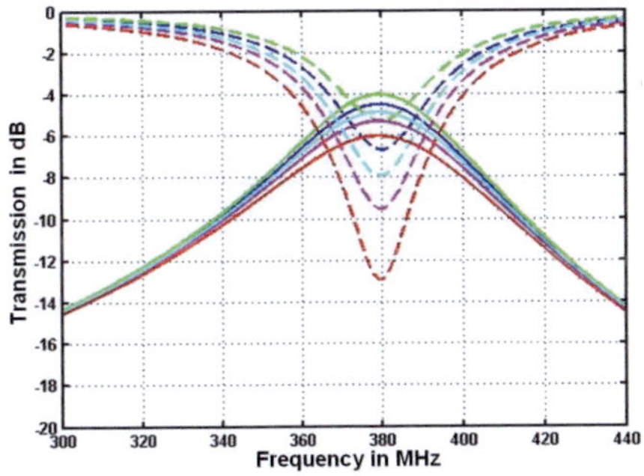

(c) ReZ = 0.5 Ω, ImZ = + j 50 Ω

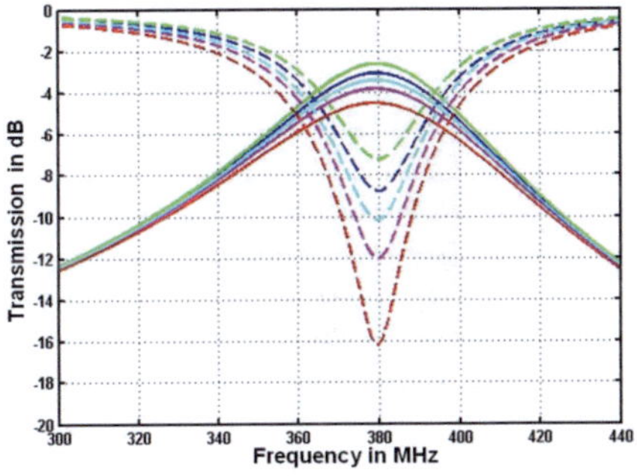

(d) ReZ = 0.8 Ω, ImZ = + j 50 Ω

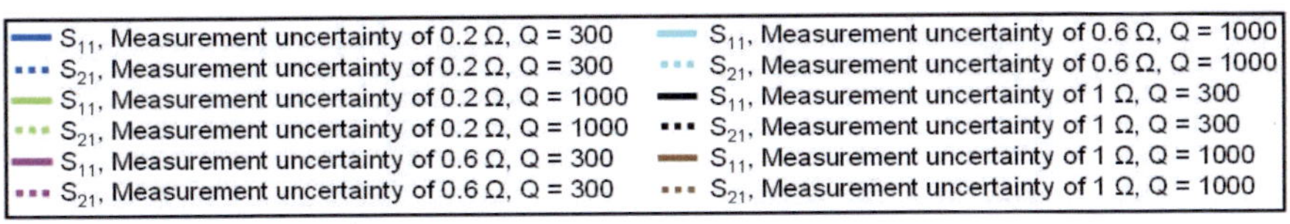

Figure 5.48: Reflection and transmission factor of the matching network of the WU with loop antenna

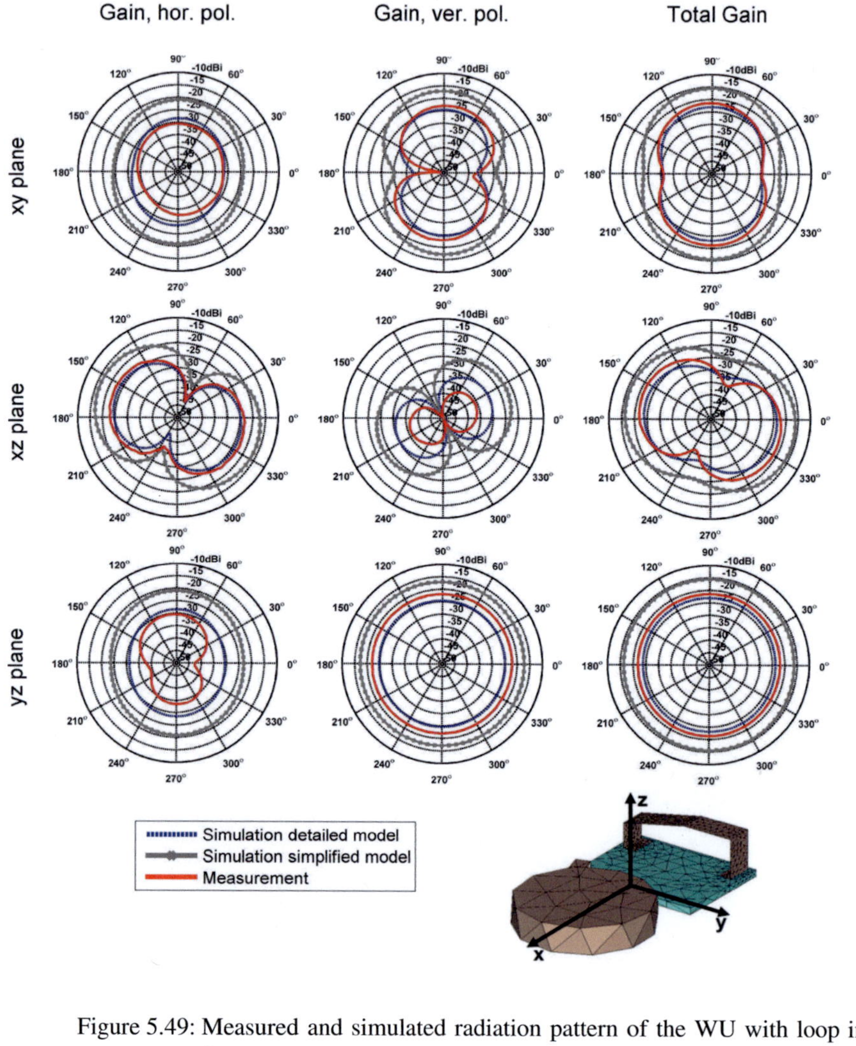

Figure 5.49: Measured and simulated radiation pattern of the WU with loop in free space

5.3.6 f Summary and Outline for the Wheel Unit with Loop Antenna

Stand alone WU with loop antenna features worse performance as the priorly analyzed stand alone WU with IFA. Nevertheless the WU will be further taken into account and inspection of the performance on the rim and with tyre will be conducted. Few properties differ the loop antenna from other WU antennas under investigation:

- Small loop antenna means relatively low efficiency in free space.
- All developed radiating elements for the presented WUs are electric antennas and the loop is the only magnetic antenna type.
- The loop features very low impedance.

Following consequences out of the upon points make the loop antenna interesting for application in TPMS:

- Dielectric surrounding should have much less influence on the loop antenna than on electric antenna types.
- The antenna performance should be more stable.

Further results are presented in Chapter 6.8.2 and Chapter 6.8.3 with the rim and tyre influence.

5.3.7 Conclusions of Wheel Unit Antenna Development

Firstly the already existing Wheel Units were analyzed. Secondly as these WUs did not satisfy the requirements, new solutions were under investigation. Fig. 5.50 outlines very well the considerations on the WU.

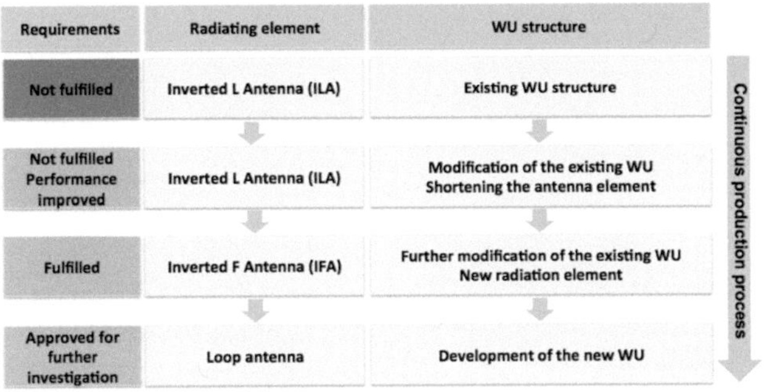

Figure 5.50: Development approach for the Wheel Unit - optimization process

Summing up, neither short nor long monopole antenna featured the expected performance. Two new possible solutions were introduced: WU equipped with an Inverted F Antenna and WU equipped with a loop antenna.

Both of the presented solutions will be further investigated with the rim and different tyre types in Chapter 6.

6 Automotive Electromagnetic Simulation

Chapter 6 deals with automotive electromagnetic simulation. The basis for the complex considerations of full automotive simulations incorporating vehicle body were laid in all previous chapters. Tyre classification (Chapter 3) took place, material parameters were analyzed (Chapter 4) and adequate antenna modules were presented and analyzed (Chapter 5). Further final work incorporates all up to this point gathered results. Automotive systems already presented in Chapter 2 will be here throughly examined with the means of different electromagnetic simulation tools. The aim is to find out to which extent an engineer may relay on results of numerical simulations in order to shorten or save time spent on practical measurements.

Firstly the most appropriate simulation tools are described. Vehicle body and tyre models are presented, followed by PASE, RKE, tyre and TPMS simulation results.

For most modern antennas, their radiation patterns are very complex so that closed-form mathematical expressions are not available. Even in cases where expressions are available their form is so complex that integration to find the radiated power required to compute the maximum directivity cannot be performed [39].

The tasks to be solved get even more complex as such structures as not only antenna geometry but also vehicle body (Chapter 6.2) with metallic as well as dielectric components have to be considered.

In consequence the problem complexity raises the necessity of multiple measurements. Performing solely the measurements there is high probability that an optimal solution will never be found. The concept is to apply numerical methods supported by high speed computer systems. Meanwhile computer simulation of electromagnetic behaviour is a vital and inseparable part of the design of electronic devices [115].

Electromagnetic simulation is the process of modelling the interaction of electromagnetic fields with physical objects and the environment. Simulation has got on purpose the reduction of effort committed to the experimental tasks. Hypothetical situations are modelled in order to study how a real system would work without the necessity of setting the real system up. By changing either simulation components (e.g. different antenna types, car bodies or tyre size) or parameter (e.g. antenna size) the behaviour of the system may be predicted. The knowledge gained in this way determines which measurements should be conducted. There is, however, one important condition: simulation reliability and conviction that the simulations reflect the reality.

Simulation accuracy highly depends on the hypotheses made for a given problem. In radio frequency techniques very important are:

- The level of system simplification.
  With respect to the wavelength of the frequency of interest it is important to simplify the detailed geometry.
- The geometric complexity scale.
  In special case of automotive electromagnetic simulation from simple and small (in respect to the wavelength) metallic radiators through antenna modules with dielectric materials up to vehicle body and complicated 3D structures such as tyres.
- Material parameters discussed in Chapter 4.
  Material parameter values have significant influence on the achievable results.

This diversity of geometric scales and sizes corresponds to different physical behaviours with consequently call for different modelling approaches (e.g. Method of Moments (MoM), Finite Element Method (FEM)) and different simulation environments (e.g. FEKO, Ansoft HFSS, Micowave Studio).

6.1 Numerical Methods

Numerical methods applied in CEM (Computational Electromagnetics) involve efficient approximations to Maxwell's equations (here in differential form):

$$\nabla \times \vec{E} = -\frac{\delta \vec{B}}{\delta t} \qquad \nabla \times \vec{H} = \vec{J} + \frac{\delta \vec{D}}{\delta t} \tag{6.1}$$

$$\nabla \cdot \vec{D} = \rho \qquad \nabla \cdot \vec{B} = 0 \tag{6.2}$$

As the solution of the above equations is very complex, for realistic problems numerical methods are required [129]. There are many CEM techniques and corresponding software tools for solving electromagnetic problems. The core hereto are the electromagentic theory and the numerical methods. Three main numerical techniques in CEM are: the Finite Difference Time Domain (FDTD) method, the Method of Moments (MoM) and the Finite Element Method (FEM). The discussion about the application of these methods in this thesis is focused solely on applications in RF engineering.

From the end user point of view, the basic operation blocks and philosophy behind each method are very similar. The problem to be solved has to be modelled with the means of specific software user interface or geometrical data has to be imported. Due to increased simulation time for complex structures it is of favour to simplify the model of interest. Independent of applied method, the submitted geometrical model is divided into a certain number of elements. This process is called discretization or meshing. The number of discrete elements is dependent on the operation frequency and simulated object size. The higher the frequency, the more elements are required and the denser the resulting mesh. The calculations, mathematical operations are conducted for each mesh element separately. The unknown for each mesh element (in MoM the current, for FEM and FDTD the electrical field) is described by space functional dependence. The here applied equations are so called basis expansion functions.

Simulation accuracy is related to the mesh quality (element size and mesh form) - the general validity, that the finer the mesh, the better accuracy, is not applicable and not true. There is always a trade-off to be met between simulation accuracy, simulation time, computational resources and mesh size. Thumb rule says that the maximum element edge length should be about one tenth of the wavelength. There is also a lower limit, boundary for the mesh size, which if violated causes long simulation times and inaccurate to wrong results.

It is important to mention that the thumb rule may not always be applied due to size of the object to be simulated and its equivalent components. In this thesis, whole range of objects with large size diversification are simulated. The mesh elements have to be much smaller than one tenth of the wavelength for example for the WU. At the frequency of 434 MHz, the wavelength is 69 cm - according to the general rules an edge of the meshed element would have to be 6.9 cm long, whereas WU overall size does not exceed 6.5 cm.

Due to the above reasons, deep understanding, experience and sensibility is required in order to set an accurate and efficient simulation model. Important is also to choose appropriate simulation method. Every of the already mentioned methods has got it's strong and weak points. The key features for MoM, FEM and FDTD are listed in Table 6.1 [117], [118], [119], [120].

Summing up, both MoM and FEM are well suited for single or few frequency points simulation, whereas FDTD method is capable of calculating broadband system behaviour in reasonable time. There is, however a restriction that FDTD method is effective for object with orthogonal boundaries. FEM is accurate for all object kinds but as these objects exceed certain dimensions in respect to the operation frequency then the simulations take long time. MoM has got no problems with big structures but precise modelling of dielectric bodies longs for deeper method understanding and experience.

None of the above methods provide good solution for all type of electromagnetic problems. Every modelling method has its merits and limitations. This is the reason why there is necessity to use different simulation tools and methods behind. Resulting from the above considerations, for the purpose of this thesis, two methods and equivalent simulation software was chosen: FEM implemented in Ansoft HFSS and MoM implemented in FEKO.

Method	FEM	MoM	FDTD
Domain	frequency	frequency	time
Meshing requirement, discretization	entire computational domain (including boundary air box) is to be meshed	only object of interest is meshed, "source method"	the whole simulation domain is discretized
Basic mesh element	tetrahedron	for conducting surfaces triangle, for dielectric either triangle or cuboid (dependent on the solution method)	cube
Boundary conditions	absorbing boundary conditions	no need for spacial boundary condition	absorbing boundary conditions
Core unknown	electric / magnetic field	current distribution	electric / magnetic field
Solution method	linear equations, sparse matrix	linear equations, full matrix	iterations
Numerical effort (N - number of unknowns)	N^2	N^3 (direct solver), N^2 (iterative solver), $NxlogN$ when matrix compression techniques implemented	N
Broadband simulation	every single frequency point needs to be solved separately	every single frequency point needs to be solved separately	capable, efficient
Complex, arbitrary shaped 3D geometries	suitable	suitable	capable
Dielectric materials	easy to model	suitable (SEP, VEP, planar Green function, thin dielectric sheet and dielectric coating method)	easy to model
Large objects	take long time	suitable	suitable when orthogonal bodies
Examples of objects to be simulated	mobile phone antenna with covers, battery, screen, etc.	metallic antenna objects, big structures, automotive antenna, antenna with radome, patch antennas	Ultra Wide Band (UWB) antennas
Simulator tool examples	Ansoft HFSS	FEKO	Empire, EM explorer, Microwave Studio

Table 6.1: Comparison of three main numerical techniques in CEM

An overview of the collocation of the simulation method in respect to the simulated object is depicted in Fig. 6.1. The dependence is presented upon object size and complexity. Single objects under test as WU, receiver or even WU on the rim were simulated with the help of HFSS. Large scale CEM problems face however memory and CPU limitations and therewith very long simulation duration. The whole TPMS, RKE and PASE systems incorporating vehicle body were simulated with MoM (FEKO). Therefore models of single system base elements had to be modelled in FEKO as well.

The simulation models for the purpose of this thesis and future investigations are built and set together in a modular way - it is very easy to exchange for example a WU antenna type or tyre in TPMS system simulation or even the vehicle body in order to experience and compare changes in the results. All geometrical models have been designed parametric. WU model is presented in Chapter 5.3, receiver model in Chapter 5.2. Tyre model as well as simulations of the whole system with the vehicle body are presented and discussed in the following chapters.

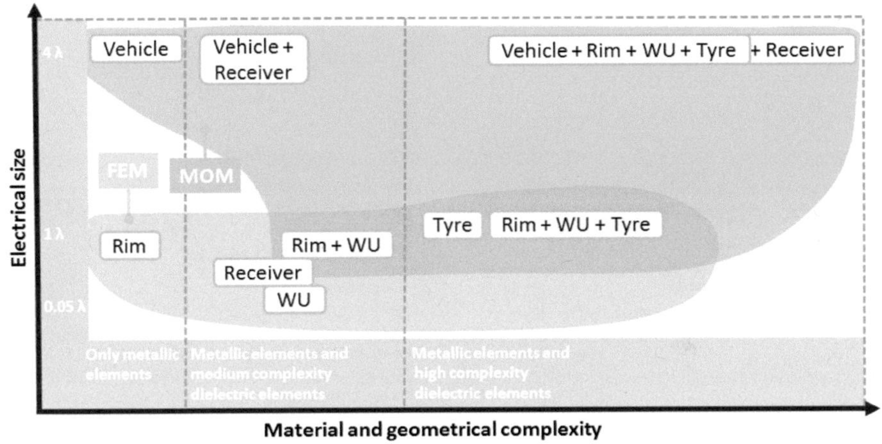

Figure 6.1: Recommendation of simulation method in dependency from object size and complexity

6.2 Vehicle Body Simulation Model

Vehicle body model is an elemental component in the automotive CEM. For simulation purposes solely metallic shell body was taken into account. The availability of the geometrical CAD (Computer Aided Design) model reduces the work load and preparation time in the simulation process.

Mechanical CAD models are however not suitable for the electromagnetic simulations. This is due to very detailed structure with many small elements like screw holes, hinges, door locks and other small radius curvature element profiles. Even if a meshing of such model would be possible then the above described body details would boost mesh element number and therewith the simulation time.

Model simplification plays a crucial role and influences the simulation performance (time as well as accuracy). All model preparation stages for the simulation purposes in this thesis are depicted in Fig. 6.2. Real vehicle is presented in Fig. 6.2a and corresponding CAD model in Fig 6.2b. The CAD model was simplified, what is visible in Fig. 6.2c. Fig. 6.2d depicts the already meshed model for electromagnetic simulation.

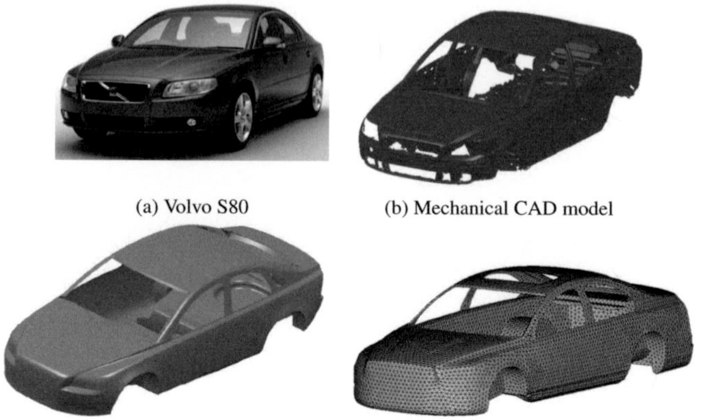

(a) Volvo S80 (b) Mechanical CAD model

(c) Simplified CAD model (d) Meshed model for simulation purposes

Figure 6.2: Vehicle simulation model

6.3 Rim Simulation Model

Rim simulation model was built with the help of FEKO graphical interface. The model, depicted in Fig. 6.3, is fully parametrized and therewith scalable. Meshed simulation model is visible in Fig.6.3c and Fig.6.3d.

Important for the accurate simulation results is the rim inner profile. The nearest surrounding of the WU should be modelled accurately as shown in Fig. 6.4. In order to reproduce any rim profile and size (w - rim width, d - rim diameter), it was parametrized. Meshed rim together with positioned WU is depicted in Fig. B.2b.

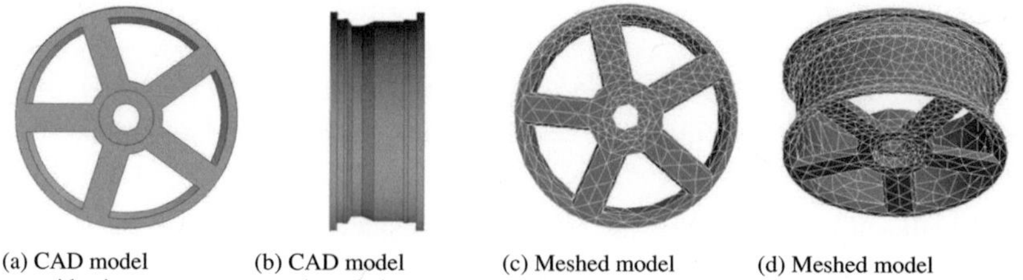

(a) CAD model - side view (b) CAD model - lateral view (c) Meshed model - side view (d) Meshed model - lateral view

Figure 6.3: Rim simulation model

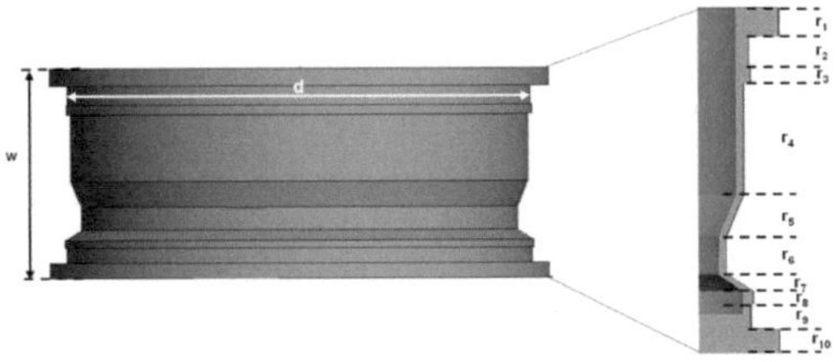

Figure 6.4: CAD rim model - profile

6.4 Tyre Simulation Model

Tyre construction was already discussed in Chapter 3.2. Material properties of slick, conventional and run flat tyre representatives were determined in Chapter 4.4, Chapter 4.5 and Chapter 4.6 respectively. The challenge right now is to introduce as accurate electromagnetic model of a layered, non infinite, curvature structure for three above mentioned tyre types. The designed models should be scalable in order to reproduce different tyres on request.

All tyres are modeled without adequate tyre tread profile - such model accuracy is not necessary for precise simulation result. An example of the tyre model (outer view) without the rim is presented in Fig. 6.5

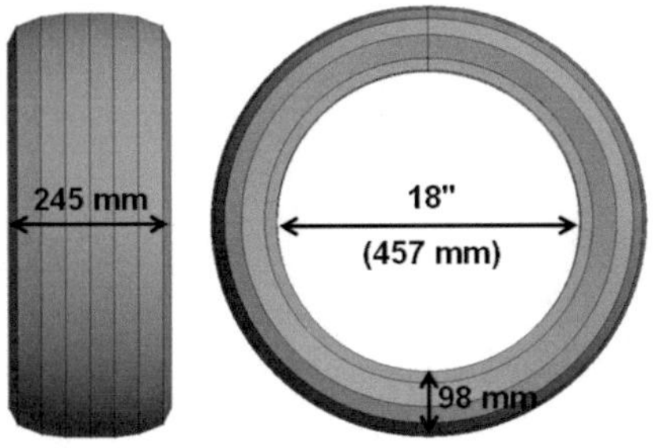

(a) Side view (b) Lateral view

Figure 6.5: Tyre simulation model - outer view (example for 245 / 55 R 18 tyre)

6.4.1 Slick Tyre

Slick tyre model due to tyre construction simplicity is the least complicated. The tyre features only two material layers without any metallic reinforcement in between (Chapter 3.1). The model of the tyre together with the rim is presented in Fig. 6.6a.

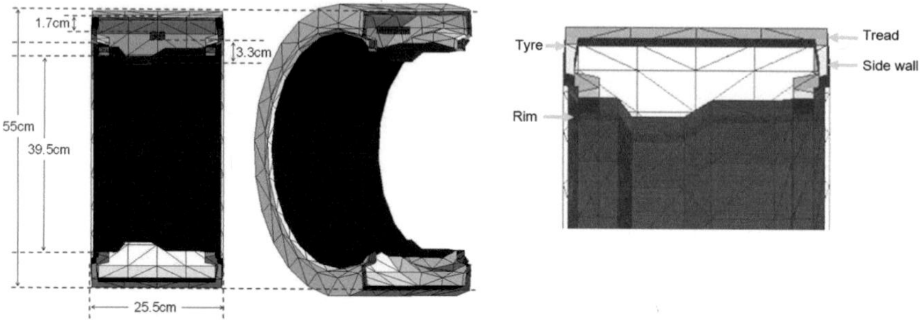

(a) Cross section (b) Indication of tyre parts

Figure 6.6: Slick tyre simulation model

For this specific tyre type it insufficient to define two rubber regions: one for the tread and another one for the sidewalls (Fig.6.6b). For both of these regions the weighted average (in respect to the material thickness) of the electrical material parameters was calculated. The values of the electrical parameters for the simulation purposes are presented in Fig. 6.7.

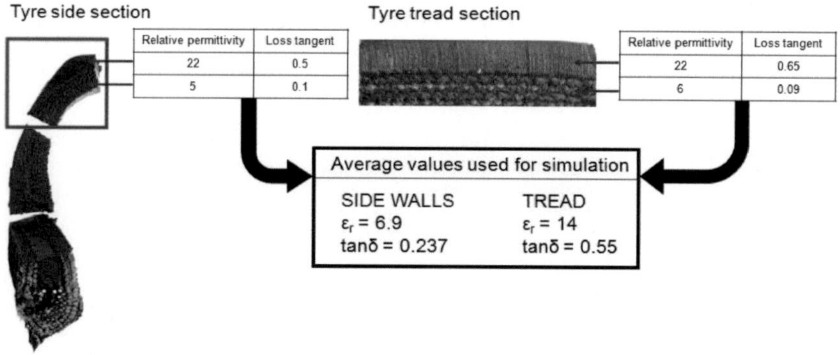

Figure 6.7: Applied average of the measured material parameters of slick tyre at 433.92 MHz at room temperature

With the CAD reproduced slick tyre geometry as well as above mentioned electrical parameters, the slick tyre model is ready to be simulated. The results may be seen in Chapter 6.8.1.

6.4.2 Conventional Tyre

For conventional tyres similar extent of simplification as for slick tyres is not possible. As presented in Chapter 3.2, tyre geometry is much more complex. All tyre layers that are taken into account for the simulation model are indicated in Fig. 6.8.

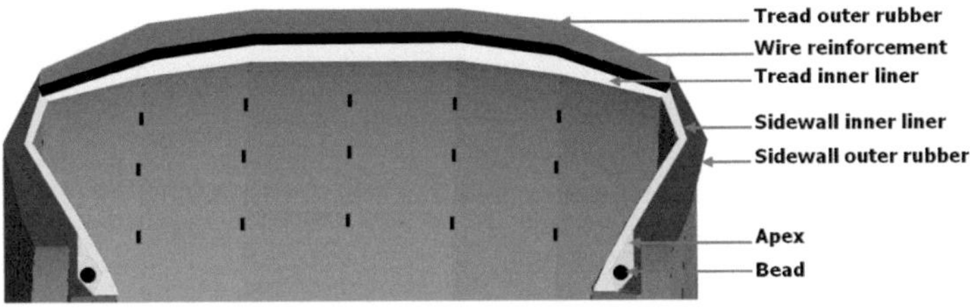

Figure 6.8: Conventional tyre simulation model - cross section

Due to metallic wire reinforcement it is not possible for this tyre type to simplify the model to only two rubber regions as for slick tyre. All rubber regions, metallic reinforcements and adequate electrical parameters have to be taken into account. The tyre model (3D representation) is depicted in Fig. 6.9.

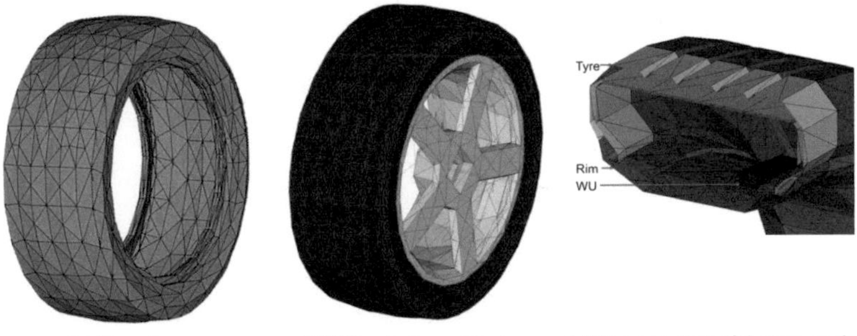

(a) Tyre alone (b) Wheel - tyre with rim (c) Cross section of the tyre on the rim with WU

Figure 6.9: Conventional tyre simulation model meshed for 433.92 MHz

The here presented model in combination with defined electrical parameter (Chapter 4.5) are the base for the wheel RF characterization together with the existing WUs. The results are presented in Chapter 6.8.3 .

6.4.3 Self Supporting Runflat Tyre

The model of the SSR tyre features one significant difference in comparison to the conventional tyre. This is the side wall reinforcement - so called extra rubber. Simulation model of the relevant tyre parts is presented in Fig. 6.10. 3D representation would look the same as for conventional tyre.

The here presented model in combination with defined electrical paremeter (Chapter 4.6) are the base for the wheel RF characterization together with the existing WUs. The results are presented in Chapter 6.8.4.

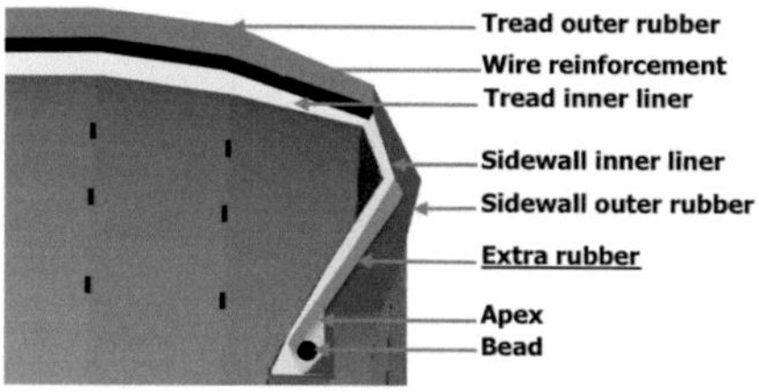

Figure 6.10: SSR tyre simulation model - cross section

6.5 Model Performance Evaluation

Performance of the prepared simulation models is evaluated upon consistency between results achieved in numerical electromagnetic simulations and equivalent, corresponding measurements.

It was already proved in Chapter 5 that the simulation models of the system base components such as a receiver and a stand alone WU are accurate. For the overall performance evaluation all vehicle relevant components are taken into account. Here the model simplification grade plays a very important role. Both mechanical (structure, built up) and electrical properties (Chapter 4) have to be considered.

For the systems under investigation the model performance evaluation is done for:

- PASE: - field distribution, expansion around the vehicle
- RKE: - characteristic of the field level around the vehicle
  - receiver radiation characteristic
- TPMS: - radiation pattern of the WU on the rim
  - radiation pattern of the WU on the rim with tyre
  - transmission parameter between the receiver and the WU for each wheel

6.6 PASE System Simulation

The functionality of the PAssive Start and Entry System is already described in Chapter 2.2. In order to design an efficient PASE system for any vehicle a reliable simulation system is of favour. With such a theoretical simulation tool the system behaviour prediction is possible for any wished antenna positioning and any vehicle body shape.

Antenna distribution within the vehicle body plays a crucial role for PASE system. The goal is to reach an illumination area at the driver's side along the vehicle body as well as at the trunk region. Many considerations were made upon the type of the antennas that should be used. The first idea was to employ the air coils; however, the laminar building area necessary in the vehicle turned out to be problematic. Coil antennas on a ferrite rod with quadratic cross section seem to be the most appropriate alternative. Among the benefits, for the proposed antennas, is the simple integration in the car body and the fact that the magnetic flux density ($\vec{B}(T)$) outgoing from the coil decreases rapidly with the distance (at the distance of one metre the magnetic field values are in range of nT).

6.6.1 Simulation Model

Three antennas are used to assure an appropriate field distribution on the driver's side and at the back of the car - depicted in Fig. 6.11.

Two of the antennas are built into the door handles at the driver's side and one additional antenna is placed in the rear bumper in order to cover the area around the trunk. The antennas are placed respectively at the height of 90 cm and 50 cm from the ground.

The employed antenna is depicted in Fig. 6.12a. Presented coil antenna has got the overall dimensions of 4 mm x 15 mm x 86 mm and the inductivity of L= 820 nH, ϵ_r = 60. The number of turns equals n = 160. Simulation model (FEKO) was formed with metal wire segments and dielectric cuboids and is depicted in Fig. 6.12b.

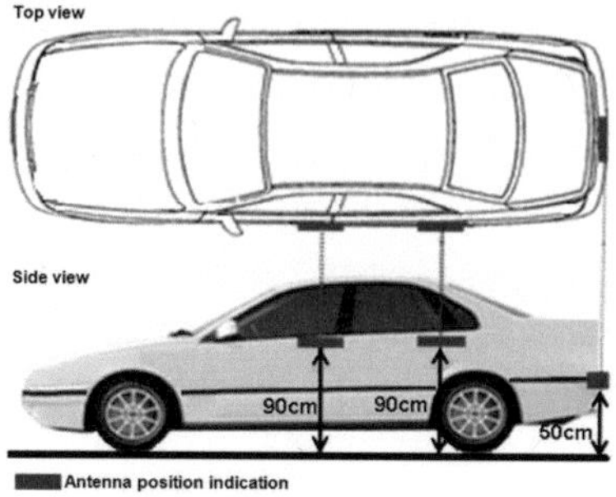

Figure 6.11: Coil antenna positioning within the vehicle

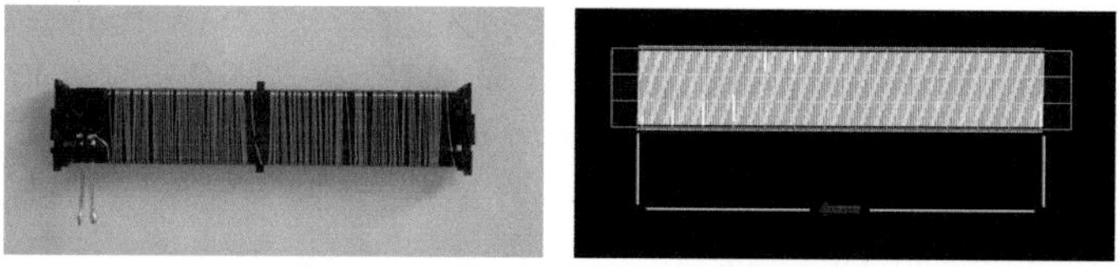

(a) Antenna sample

(b) FEKO simulation model

Figure 6.12: Coil antenna for PASE system

6.6.2 Simulated and Measured Field Distribution

Underneath different antenna positions will be simulated. It will be examined whether the employed antennas assure communication with the access card. The sensitivity of the typical access card in the production ranges between 3 nT and 4 nT and therefore the boundaries of these field values were observed. The access card that was used during all measurements had the sensitivity of 3.2 nT.

6.6.2 a One Antenna at the Vehicle Side

Primarily, the case of one antenna on the car side was examined. The task was to check whether only one antenna would be enough to assure the appropriate key reception range at the car side. Two different positions were tested: in the first case the antenna was built in driver's door handle and in the second case at the passenger's door handle on the left side of the vehicle. In both attempts for the simulation purposes the antenna was placed 90 cm above the ground and very near to the vehicle body, at the distance of 30 mm. The antenna was supplied with the current of 170 mA.

For the antenna placed at the driver's door, the view of the car cross section and the magnetic field is presented in Fig. 6.13a, for the antenna at the back door in Fig. 6.13b.

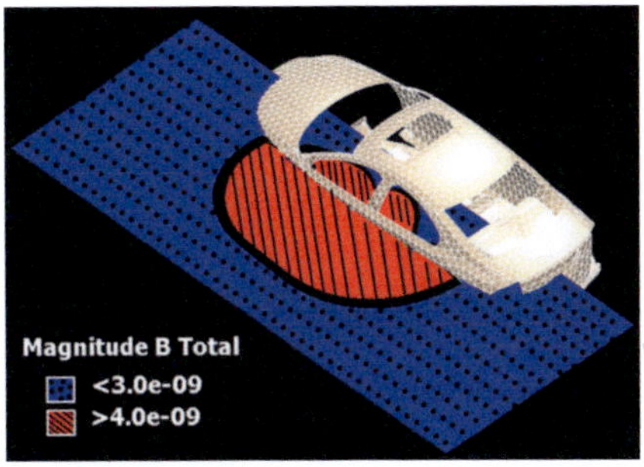

(a) FEKO simulation - Antenna at the driver's door

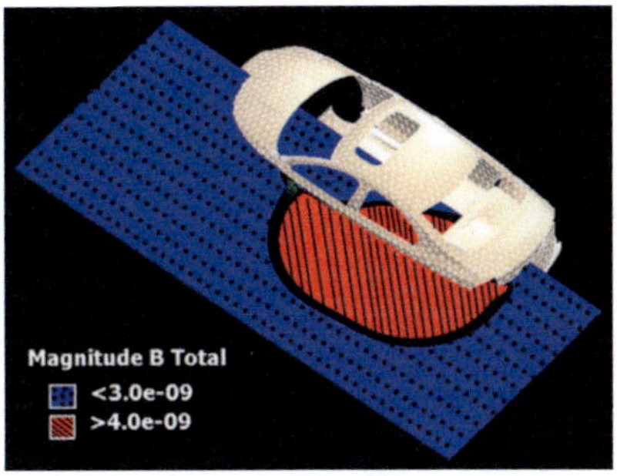

(b) FEKO simulation - Antenna at the back left door

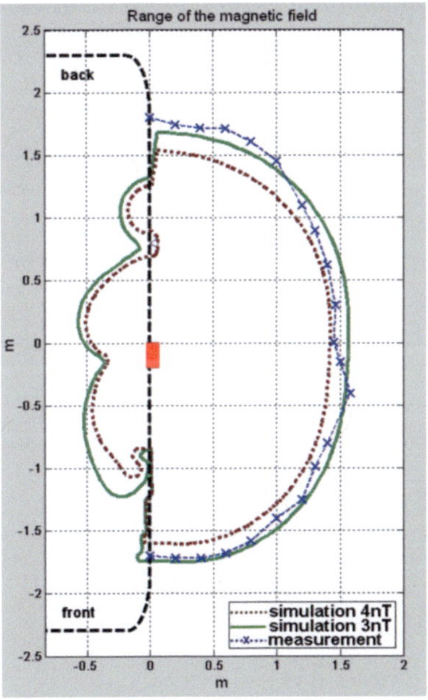

(c) Measurement and simulation comparison - antenna at the driver's door

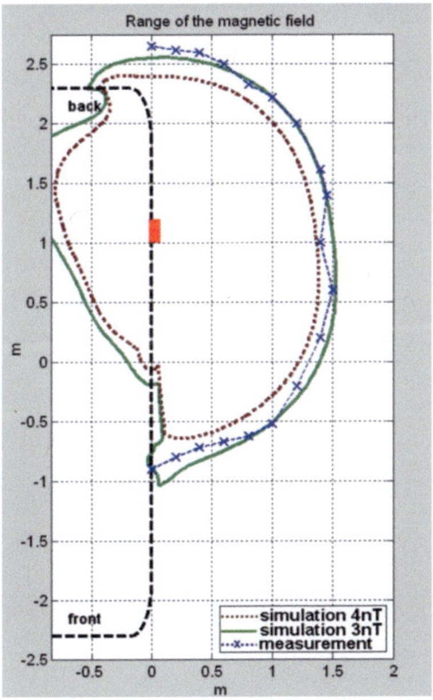

(d) Measurement and simulation comparison - antenna at the back door

Figure 6.13: Simulated and measured magnetic field distribution for one antenna placed at the driver's and at the back door

Different colour marking corresponds to the values of 4 nT and 3 nT respectively. The black transient area represents the values greater than 3 nT and smaller 4 nT. It is significant that for the field lower than 3 nT there is no access card recognition possible (area marked blue).

For verification of the above simulated, theoretical results, adequate measurements have been performed - the hereto results are presented in Fig. 6.13c and Fig. 6.13d. The car body has been depicted with black dashed line with indication of the orientation (front and back). The red rectangle indicates the antenna, the green solid line and the brown dashed line the magnetic filed values of 3 nT and 4 nT respectively. The measurements (x markers) verified the range predicted with the simulations. In the perpendicular direction away from the car the maximum range is 1.5 m and along the car 1.7 m in both directions away from the antenna.

Concerning the maximum range of the antenna in perpendicular direction to the car body, there's no difference whether the antenna is placed at the front or the back door. However from the above, it is clear that only one antenna at the car side cannot guarantee the reception of the access card in all places along the car side. This fact constraints the use of two antennas at the car side.

6.6.2 b Two Antennas at the Vehicle Side

When using two antennas, considerations about the mutual influence had to be made. Most important was the applied current flow direction. First of all two antennas free in space were simulated, both with common and differential excitation, respectively Fig. 6.14a and Fig. 6.14b.

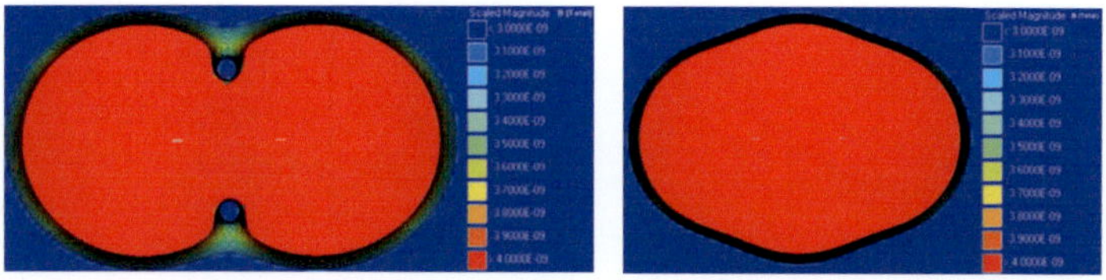

(a) Common excitation (b) Differential excitation

Figure 6.14: Magnetic field distribution of two aligned ferrite rod antennas

In both cases the antennas were placed in the distance of 83 cm from each other, just as if built into the door handles. Fig. 6.14a and Fig. 6.14b depict magnetic flux density. When the antennas are fed in phase two minima occur, whereas the antennas with differential excitation are not affricated with any strong field fluctuations. Therefore antennas with differential feeding were then assembled in the car handles.

Both antennas were fed with the same current amplitude ($I_1 = I_2 = 170$ mA). Antenna arrangement is depicted in Fig. 6.15a. Simulated magnetic flux density is shown in Fig. 6.15b. As depicted in Fig. 6.15c - the range of the magnetic field resulting from two antennas is the superposition of the fields resulted by single antennas. The maximum range in the perpendicular direction is over 1.6 m and in the longitude direction the access card is recognized both in the area around the front and the back door. Therewith, the requirement for the good limitation of the access card recognition is fulfilled. Only if the owner is directly approaching the vehicle door the access card is recognized.

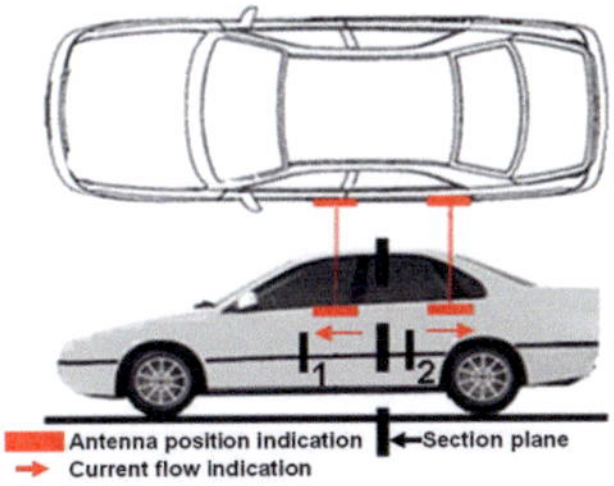

(a) Antenna orientation

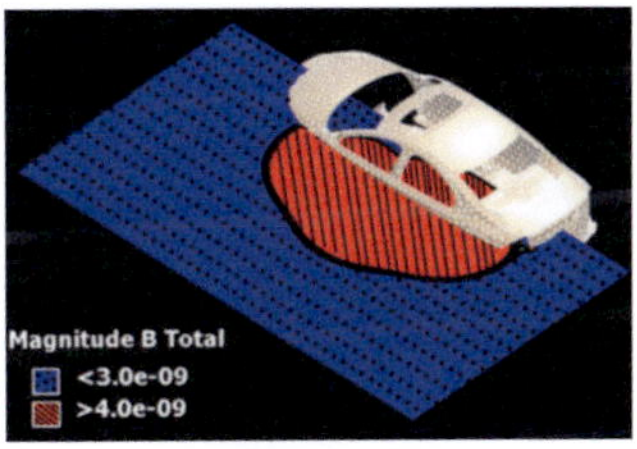

(b) FEKO simulation

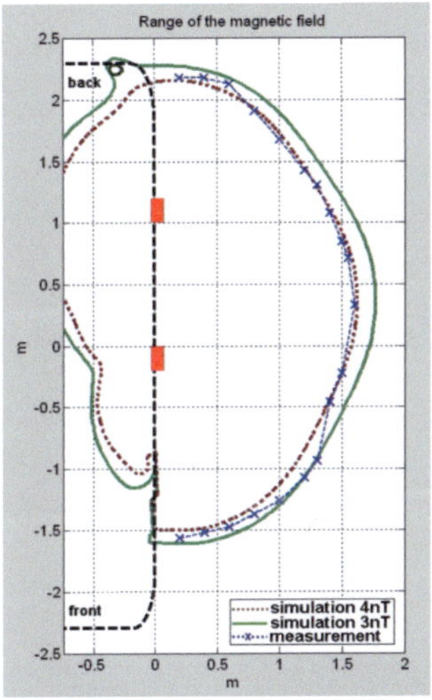

(c) Comparison of measured and simulated values

Figure 6.15: Simulated and measured magnetic field distribution for two antennas with differential excitation placed at the vehicle side

One point is still open, namely where does the user have to put his access card. Will it be recognized when carried in the pocket, in the bag on the shoulder or in a trolley somewhere near the ground?

The field distribution in the vertical cut plane is shown in Fig. 6.16a. The distribution of the magnetic field was read exactly in the middle of the vehicle between two positioned antennas. As depicted in Fig. 6.16b and confirmed with the appropriate measurement the key is recognized at practically any height. At the distance of 1.5m the key is recognized up to the height of 1.5 m and when approaching closer to the car even to the height of 2.3 m.

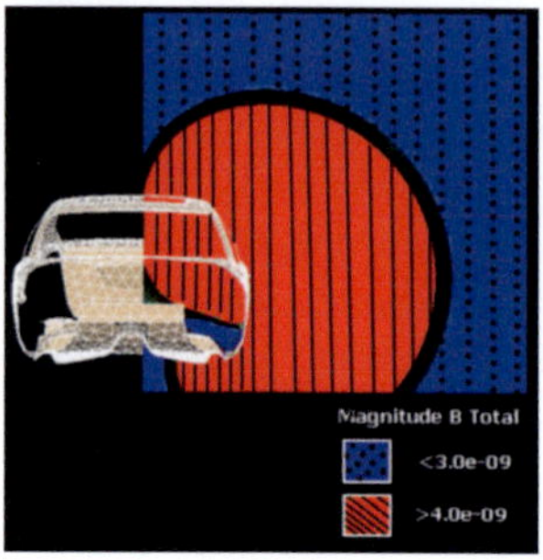

(a) FEKO simulation

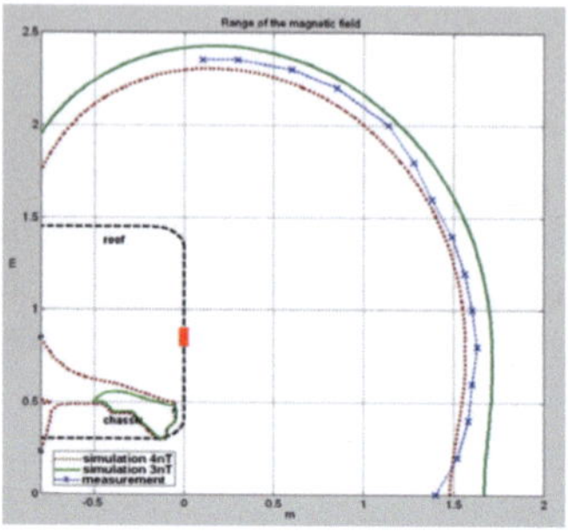

(b) Measurement and simulation comparison

Figure 6.16: Magnetic field distribution of two aligned ferrite rod antennas - vertical cross section

6.6.2 c One Antenna Placed in the Rear Bumper

To assure the comfortable access to the trunk one antenna was positioned in the rear bumper. The antenna as depicted in Fig. 6.11 was placed at the height of 50 cm from the ground, the applied current was 270 mA. In Fig. 6.17a the magnetic field distribution is depicted.

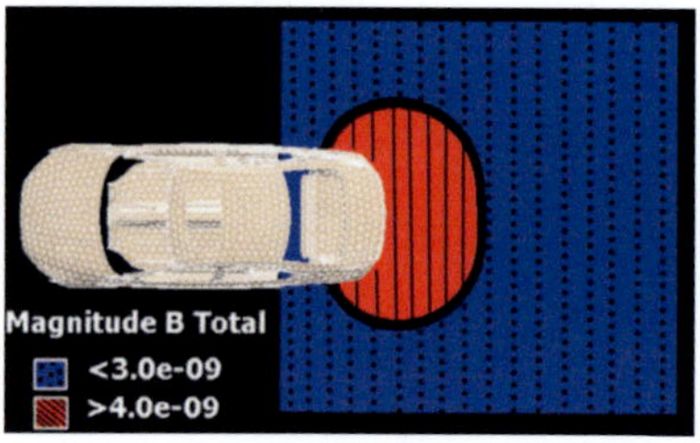

(a) FEKO Simulation

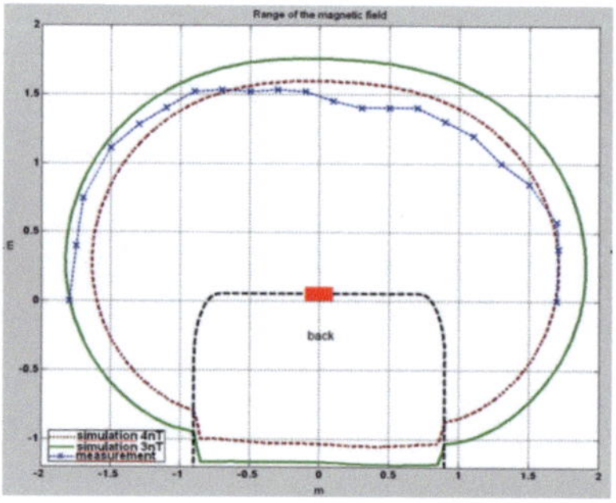

(b) Measurement and simulation comparison

Figure 6.17: Magnetic field distribution of one ferrite rod antenna placed in the rear bumper

6.6.2 d Conclusions

The executed simulations allow predicting the performance of the vehicle card access system. The measurements show that the performed simulations are very accurate - the small differences between the simulated and measured values are negligible, they do not exceed 10 cm. The reason for slight differences may be the discrepancy between the electrical properties of the real ground (asphalt) in comparison with the simulated perfect electric conducting plane as well as simplification of the vehicle body.

6.7 RKE Simulation

Remote Keyless Entry system was already introduced in Chapter 2.1 and the basic system architecture was described in Chapter 2.4. It is now time for a closer consideration on RKE system evaluation, optimum achievable results and system simulation possibilities.

Crucial point is to guarantee good communication between the key fob and the in-vehicle installed receiver module for a predefined distance d (Fig. 6.18).

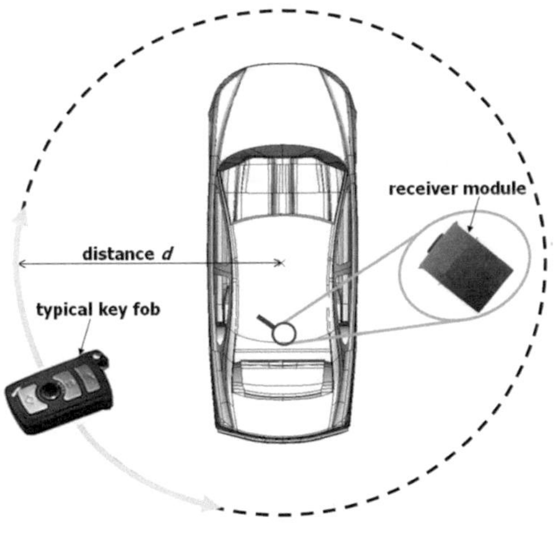

Figure 6.18: Main components of RKE system

The maximum reception distance is not only dependent on the characteristics of the integrated antennas in the key fob (Chapter 5.3) but as well as in the receiver module (Chapter 5.2). Decisive role plays the body of the vehicle as well as the characteristic of the surroundings: type of ground, free space area or in city array with buildings, trees and other vehicles and objects. Therefore, it is necessary to characterize the radio frequency propagation channel and possible ideal propagation conditions.

6.7.1 Wave Propagation Schematics - Problem Simplification

Wave propagation model for a typical RKE system is depicted in Fig. 6.19. As the receiver depicts both the directly propagating wave and the wave reflected from the ground, the received power P_r at the antenna feeding point in the receiver depends on the vehicle and its nearest environment. The ground is playing here a particularly important role in addition to system and link parameters as the transmitter power P_t, the distance d between the key and the receiver, the position (height h_t) and orientation of the key and h_r of the receiver. Different positions of the receiver within the car body are possible - it was assumed that h_r may vary between 0.3 m and 1.8 m for most vehicles and typically the key fob held in the hand would be placed at h_t of 1 m. In order to simplify the problem, make it independent

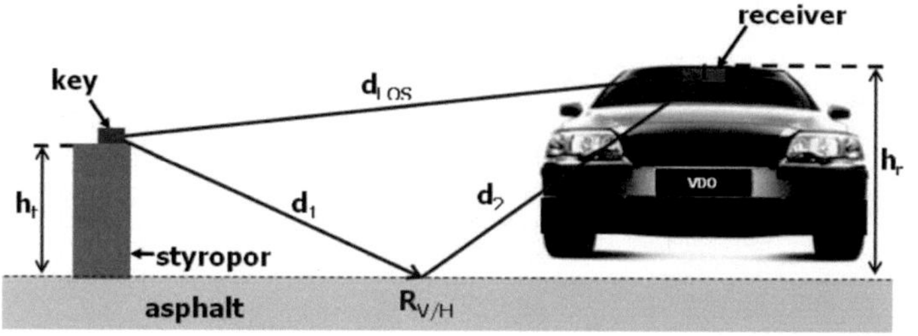

Figure 6.19: Wave propagation schematic for RKE system

on the vehicle body, the above scenario from Fig. 6.19 was isolated. From now on no vehicle body will be considered but only transmission between two stand alone horizontally and vertically polarised antennas. This simplified model is depicted in Fig. 6.20.

Further not only the schematics geometry but also the frequency of operation has to be considered. In the thesis discussed frequencies are 433.92 MHz and 868 MHz with wavelengths respectively of 69.1 cm and 34.5 cm.

The assumption of a flat surface is made. This assumption is not very far away from the car real environment - case of a free parking slot or an empty street. The characteristic values that could influence the wave propagation are the ground conductivity σ and the relative electric permittivity ϵ_r. Typical values for asphalt and cement are listed in Table 6.2 [124]. With the given model and the above assumption, transmission coefficient simulation was

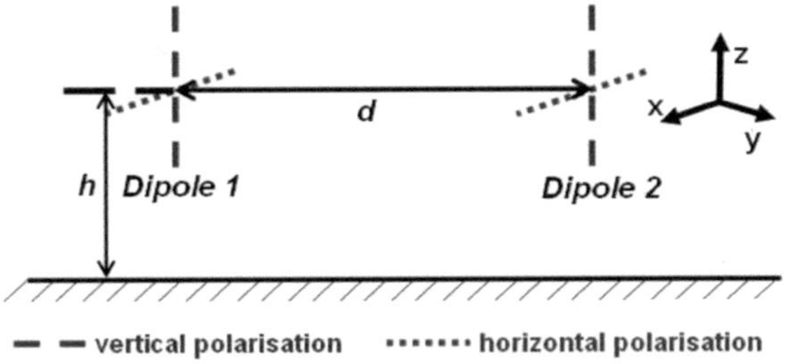

Figure 6.20: Simplified simulation model - evaluation of the transmission coefficient

Material	ϵ_r	σ in mS/m
Asphalt	2 - 12	1-100
Cement	4 - 20	1-100

Table 6.2: Electrical properties of the ground - typical values

possible. Simulations were conducted for 433.92 MHz and 868 MHz. The results are depicted in Fig. 6.21 and Fig. 6.22. For the RKE system, the distance of interest is limited to 100 m.

Attenuation values for horizontal and vertical polarization for asphalt and PEC as well as free space were under examination. The reason for such comparison lies in the inherent mathematical property of electromagnetic simulation methods and therefore tools such as FEKO. If the ground is non dielectric, the simulation duration is noteworthy decreased. The here made considerations should clarify whether it is possible and under which assumptions to carry all simulations on PEC ground in a reduced distance around the vehicle.

For the horizontal polarization, nearly no difference exists between a PEC ground and asphalt as shown in Fig. 6.21a and Fig. 6.22a. Similarly, there are also no differences between calculations based on material parameters (conductivity, relative permittivity) of asphalt. In comparison to free space, serious differences occur particularly at big distances. At a distance d = 100 m, the difference to free space amounts 15 dB at 433.92 MHz and 9 dB at 868 MHz. The frequency doesn't play a role in the presence of PEC or asphalt for a distance greater than 30 m. The impact of the frequency is noticeable in a distance smaller than 30 m, particularly in the number and location of the reflection points.

For the vertical polarization as depicted in Fig. 6.21b and Fig. 6.22b.

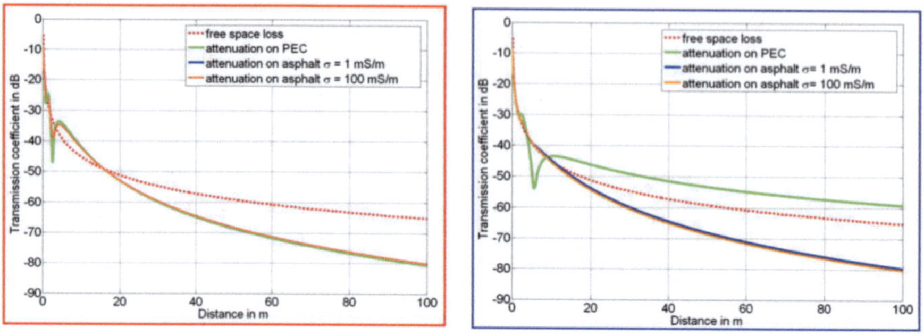

(a) Horizontal polarization

(b) Vertikal polarization

Figure 6.21: Simulated transmission coefficient for 433.92 MHz

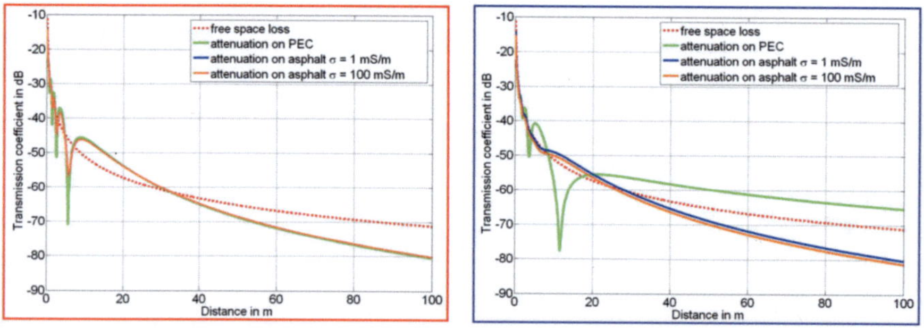

(a) Horizontal polarization

(b) Vertikal polarization

Figure 6.22: Simulated transmission coefficient for 868 MHz

Denotative difference in attenuation values can be seen for PEC and asphalt. This difference amounts 21 dB and 16 dB at $d = 100$ m respectively at 433.92 MHz and 868 MHz. Also in this case, the material parameters of the asphalt lead to negligible differences after variation of the parameter values. Furthermore, differences to free space can be clearly seen. A further evaluation of the figures for distance above 30 m shows that 10 dB are needed for doubling the range.

6.7.2 Analytical Formulation and Measurement

In order to revise the simulated values, some theoretical considerations were made. First, the Free Space Transmission factor (FSL) for both frequencies was calculated according to Eq. 6.3.

$$FSL = 20\log_{10}\left(\frac{\lambda_0}{4\cdot\pi\cdot d}\right) \tag{6.3}$$

$$FSL_{434MHz} = 20\log_{10}\left(\frac{0.69\text{ m}}{4\cdot\pi\cdot d}\right) = \begin{cases} -45\text{ dB} & \text{for } d = 10\text{ m} \\ -59\text{ dB} & \text{for } d = 50\text{ m} \\ -64\text{ dB} & \text{for } d = 90\text{ m} \end{cases} \tag{6.4}$$

$$FSL_{868MHz} = 20\log_{10}\left(\frac{0.345\text{ m}}{4\cdot\pi\cdot d}\right) = \begin{cases} -51\text{ dB} & \text{for } d = 10\text{ m} \\ -65\text{ dB} & \text{for } d = 50\text{ m} \\ -70\text{ dB} & \text{for } d = 90\text{ m} \end{cases} \tag{6.5}$$

Results for both frequencies in Eq. 6.4 and Eq. 6.5 match very well with the simulated values (Chapter 6.7.1). Further conclusion is that doubling the frequency the attenuation in free space increases by 6 dB.

The wave propagation schematics from Fig. 6.19 for real environment with the ground has to be extended by an additional contribution of the reflected wave, which is schematically depicted in Fig. 6.23. The received power $P_{r(V/H)}$ can be expressed with Eq. 6.7 [125] , where the indices "V/H" indicate the vertical and horizontal polarization.

$$P_{r_{V/H}} = \left(\frac{\lambda_0}{4\cdot\pi}\right)^2 \cdot G_r \cdot G_t \cdot P_t \cdot A \tag{6.6}$$

$$A = \left| \underbrace{\frac{e^{-j\cdot k_0\cdot d_{LOS}}}{d_{LOS}}}_{\text{LOS component}} + \underbrace{R_{V/H}\cdot\frac{e^{-j\cdot k_0\cdot d}}{d}}_{\text{reflected component}} \right|^2$$

Where G_r is the receiver antenna gain, G_t is the transmitting antenna gain, P_t is the transmitted power, $R_{V/H}$ is the reflection factor and d is the distance travelled by the reflected wave $d = d_1 + d_2$.

$$d = \sqrt{(x_r - x_t)^2 + (h_r + h_t)^2} \tag{6.7}$$

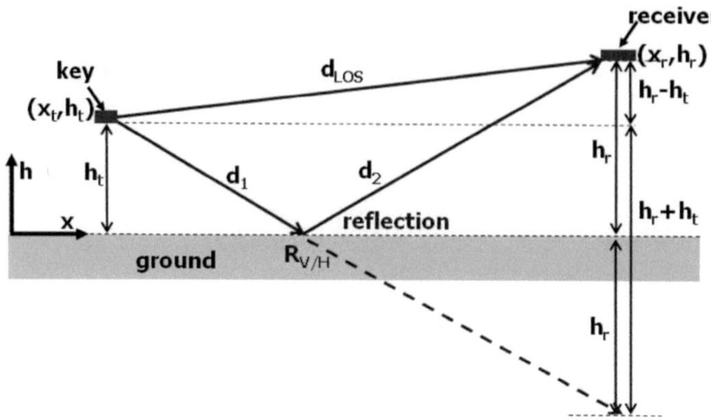

Figure 6.23: Wave propagation geometry

$$d_{LOS} = \sqrt{(x_r - x_t)^2 + (h_r - h_t)^2} \tag{6.8}$$

With the assumption of Perfect Electric Conducting ground: R_v = 1 und R_h = -1, Path Loss (PL) between the receiver and the transmitter can be expressed with Eq. 6.9 (unity antenna gain is considered).

$$PL_{PEC} = \frac{P_r}{P_t} = \begin{cases} \left(\frac{\lambda_0}{4 \cdot \pi \cdot d}\right)^2 \cdot 4 & \text{for vertical polarization} \\ \frac{h_r \cdot h_t}{d^4} & \text{for horizontal polarization} \end{cases} \tag{6.9}$$

For the case of the transmitter and receiver antenna positioned at the same height h_t = h_r = 1 m, following calculations may be made:

$$PL_{PEC_{868MHz}} = \begin{cases} -\ 59\ \text{dB} & \text{for } d = 50\ \text{m , ver. pol.} \\ -\ 67.95\ \text{dB} & \text{for } d = 50\ \text{m , hor. pol.} \end{cases} \tag{6.10}$$

$$PL_{PEC_{434MHz}} = \begin{cases} -\ 53\ \text{dB} & \text{for } d = 50\ \text{m , ver. pol.} \\ -\ 67.95\ \text{dB} & \text{for } d = 50\ \text{m , hor. pol.} \end{cases} \tag{6.11}$$

Based on Fig. 6.21 and Fig. 6.22, the transmission factor can be calculated with Eq. 6.12 for asphalt and for a distance greater than 20 m. This equation is also given by [126] for distances d»h.

$$PL_{asphalt} = -40\log_{10}\left(\frac{h}{d}\right) \tag{6.12}$$

$$PL_{asphalt} = \begin{cases} -70 \text{ dB} & \text{for } d = 50 \text{ m} \\ -83 \text{ dB} & \text{for } d = 100 \text{ m} \end{cases} \tag{6.13}$$

The theoretical estimation of the Path Loss both for free space as well as for ground (PEC and asphalt) matches very well the simulated values.

Additionally attenuation values were measured on asphalt for the frequency of 868 MHz. The comparison of the measured and simulated values is displayed in Table 6.3.

It may be observed that the discrepancy between measured and simulated values is slight and within measurement accuracy. The measurement accuracy is influenced by the non ideal measurement test area. Large test areas free of any obstacles are very rare. Most test sites are not completely free from such vegetation and construction obstacles which cause for example additional reflections.

d in m	Attenuation in dB - Measurement	Attenuation in dB - Simulation
20	55	53
40	62	62
60	71	68
80	74	74
100	78	78

Table 6.3: Comparison of measured and calculated attenuation values on asphalt

6.7.3 RKE Link Budget

Budget in economic understanding provides a forecast of revenues and expenditures. This simple explanation may be applied also when talking about link between antennas. Link budget determines whether and at which distance communication (link) between the transmitter (key fob) and receiver is possible. For the calculations antenna parameters as well environmental factors are taken into account. Link Budget for RKE may be expressed with following equation:

$$S_r \geq P_t + G_{key} + G_{car} + A + F_e + F_s \tag{6.14}$$

Revenue in RKE Link Budget is the existence of the RF link between the receiver (vehicle) and the key in RKE system.

Therewith the receiver system sensitivity S_r has to be in a certain balance, at least equal or in the optimal case greater than following listed factors:

- The transmitted power at the feeding point of the key antenna P_t. The limit for transmitted power is regulated and defined for each country.
- The average gain of the key antenna G_{key}.
- The average gain of the receiver together with the vehicle influence G_{car}.
- The attenuation A of the transmitted signal (see Chapter 6.7.2).
- The environment factor F_e, which includes the influence of the terrain relief, construction in low and high-density areas and vegetation. The assumed F_e value results from measurement and simulation comparison of attenuation values and equals 3 dB for the frequency of 868 MHz (see Table 6.3).
- The reliability system factor F_s, which comprises tolerance of all system components and ensures confidence margin for the system functionality as well as human influence.

It is important to stress that the human influence on the key antenna parameters for the range measurements is usually not taken into account - it is especially omitted in the simulation process. During the measurements a key is usually placed at a certain height from the ground on a foamed polystyrene block as in Fig. 6.19. This is for the reason that experience showed that the measurement results differ a lot dependent on who is holding the key and how. Also in real situations when a driver approaches the vehicle, the key is held in different ways or even sometimes is activated when being in a pocket. Placing a key in a controlled environment assures measurement repeatability and ensures the possibility to check the system for two key orientations that correspond to horizontal and vertical polarization.

Considering Eq. 6.14, the link between the key and the vehicle is so long guaranteed as the limit expressed by system sensitivity is not compensated by the listed factors. The in the key transmitted power P_t is known, the key gain G_{key} hand the receiver-vehicle gain G_{car} is to be measured, reliability and environmental factors F_s, F_s are empirical values based on the experience. With all given values the task is to see with the known attenuation to check at which distance from the vehicle the connection is still guaranteed - the largest possible distance.

For the general estimations an average values for the key as well as receiver-vehicle gain is considered. Parameters describing the system under investigation at 868 MHz are gathered in the underneath Table 6.4.

S_r	–115 dBm
G_{car}	0 dBi (see Chapter 5, Fig. 5.16)
G_{key}	-3 dBi
P_t	14 dBm
F_e	3 dB
F_s	0 dB
A	distance dependent

Table 6.4: List of parameters for RKE Link Budget calculation

With the above given values the RF Link Budget equation would be expressed:

$$-115\ \text{dBm} \geq \begin{cases} 14\ \text{dBm} - 3\ \text{dBi} - 3\ \text{dB} - 53\ \text{dB} & \text{for } d = 20\ \text{m} \\ 14\ \text{dBm} - 3\ \text{dBi} - 3\ \text{dB} - 62\ \text{dB} & \text{for } d = 40\ \text{m} \end{cases} \tag{6.15}$$

$$-115\ \text{dBm} \geq \begin{cases} -45\ \text{dB} & \text{for } d = 20\ \text{m} \\ -54\ \text{dB} & \text{for } d = 40\ \text{m} \end{cases} \tag{6.16}$$

With the given values link between the key and the vehicle is guaranteed for 20 m and 40 m. It is very easy to estimate from which distance the connection would not be assured, with the given values theoretically even at 100 m distance the connection should be possible.

The above calculation procedure allows to predict the RKE system functionality - at least the optimum case and theoretical distance limit.

6.7.4 RKE Link Budget Dependency on Direction

In reality neither the key fob antenna nor the vehicle with the receiver feature an isotropic radiation pattern. The key radiation pattern will not be a subject of the following discussion as the vehicle determines the system much stronger.

Upon such simulated radiation pattern it is possible to predict exactly RKE system functionality for each polarization. Moreover knowing the difference between the Attenuation values for PEC and asphalt, it is sufficient to simulate the radiation pattern of the vehicle in the nearest distance in far field region. Upon this measurement or simulation, system functionality for further distances may be estimated.

6.7.5 Conclusions

One of the most important benchmark for automobile access systems as RKE is the maximum range. The functionality of the system in terms of the distance is determined by a link budget calculation. The behaviour of the attenuation is necessary for range prediction. After comparison of the measured, simulated and analytically calculated values for the attenuation in free space and over the ground, the following conclusions can be written:

- For PEC, the difference between the horizontal and vertical polarization is particularly remarkable.
- For asphalt, there is no difference in the attenuation for both polarizations. So the attenuation is here polarization independent.
- The material parameters for asphalt do not lead to important differences between the attenuation values. This conclusion is very important for future RKE measurements. No special area test site is necessary and the measurements may be done and compared with each other independent on the asphalt type.
- The attenuation on asphalt is almost frequency independent from 30 m distance.
- In free space, 6 dB are needed to double the range. On asphalt 10 dB are necessary.

6.8 TPMS Component Simulation

In order to establish an accurate system simulation it is mandatory to assure that very simulation element is modeled in the right manner and delivers stable as well as accurate results. Complex system elements contribute to overall TPMS simulation process. The very basic components as stand alone WU and receiver were under investigation in Chapter 5. Next step is to estimate the rim and tyre influence on the WU performance. The functionality of TPMS system and therewith also its components was determined within this thesis only for frequency of 433.92 MHz.

Firstly a slick tyre with relatively uncomplicated structure was under investigation. After assuring that the simple model delivers good performance in therms of resemblance between simulated and measured characteristics, more complex tyres models (conventional, SSR) were prepared. The performance was checked always for the WU on the rim and afterwards in tyre cavity. Chapter 6.8.1 to Chapter 6.9.4 present the results of the here described investigations.

6.8.1 Slick Tyre

Measured and simulated results for radiation pattern of the chosen slick tyre are presented in Fig. 6.25. Relevant measurement plains are explained in Fig. 6.24.

(a) Sim. model - cross section (b) xy meas. plane (c) xz meas. plane (d) yz meas. plane

Figure 6.24: Orientation of slick tyre during radiation pattern measurement

The slick tyre was measured with WU equipped with the Inverted F Antenna. The influence of tyre and rim on the characteristics of the radiation pattern may be here easily observed. The rim presence causes gain losses of approx. 4 dB (with small deviations in dependency on the measurement plane and polarization). Very significant for slick tyre is the fact that the radiation pattern is not much influenced by the tyre itself. Only small changes in the amplitude and radiation pattern form are observable. This is due to the fact that the distance to the tyre thread from the WU is significant and the tyre is not reinforced by metallic wirings in the tread.

It may be easily recognized that the results for the WU on the rim as well as results in tyre cavity match well taking into account simulated and measured characteristics of the radiation pattern. Small discrepancies result from the positioning misalignment on the rotation table in the anechoic chamber. Comparison of the measured and simulated values confirms very good model accuracy.

After the above finding (accurate modeling of a relatively simple structured slick tyre) a further step may be approached and tyres with more complicated structure and their models may undergo an examination. Firstly the results for both sensors equipped with IFA and loop antenna on a rim will be discussed. Further the discussion will lead to the examples of the interaction between the conventional and SSR tyre models with both WUs.

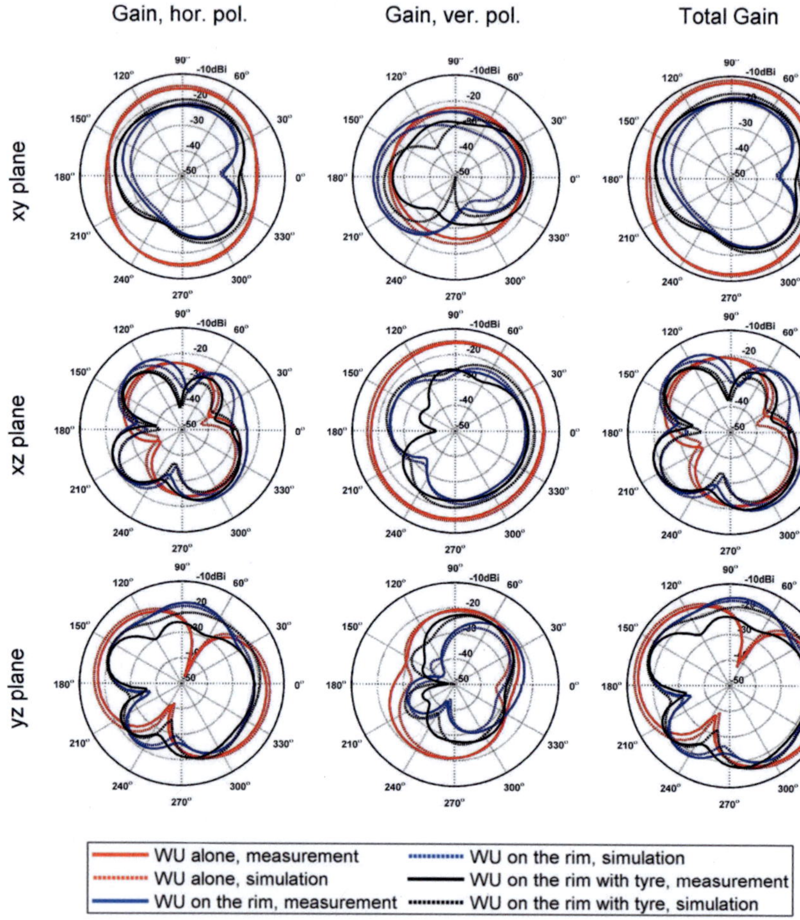

Figure 6.25: Measured and simulated radiation pattern of WU with Inverted F Antenna in slick tyre cavity

6.8.2 WU on the Rim

For the simulation purposes a fully assembled WU together with valve was positioned on two rims of choice. The reason for both measurements was to assure model accuracy and result repeatability.

For the investigations a rim with the diameter of 15" (rim equivalent for 195/65 R15 tyre, further called small rim) and a rim with the diameter of 18" (rim equivalent for 245/40 R18 tyre, further called big rim) were selected. The most important dimensions of the rim as the radius, width, number of spokes and their width, the inner and outer profile were measured and modeled exactly to correspond well with the reality. Scalable rim model was presented in Chapter 6.3.

WU on the rim for simulation and measurement purposes was oriented as in Fig. B.2. This is the reference positioning for all following radiation pattern characterization. Influence of the rim on radiation characteristics for both WUs (IFA and loop) was determined and is depicted in Fig. 6.26 for the small rim and in Fig. 6.27 for the big rim.
Comparing the above figures, the following conclusions may be drawn:

- Both rims influence the radiation pattern characteristics.
- Both shape and amplitude of the radiation pattern changed for WU with IFA and WU with loop (in comparison to the stand alone WUs).
- The radiation pattern shape for both WU with IFA and WU with loop does not change much in dependency on the rim size.
- In case of both WUs the simulation results match very well with the measurement.

Apart from the good accuracy of the achieved results, one very interesting effect may be observed. In comparison the the stand alone WUs:

- Interaction of the rim with WU antenna IFA results in slight Gain decrease.
- Interaction of the rim with WU antenna loop results in Gain increase.

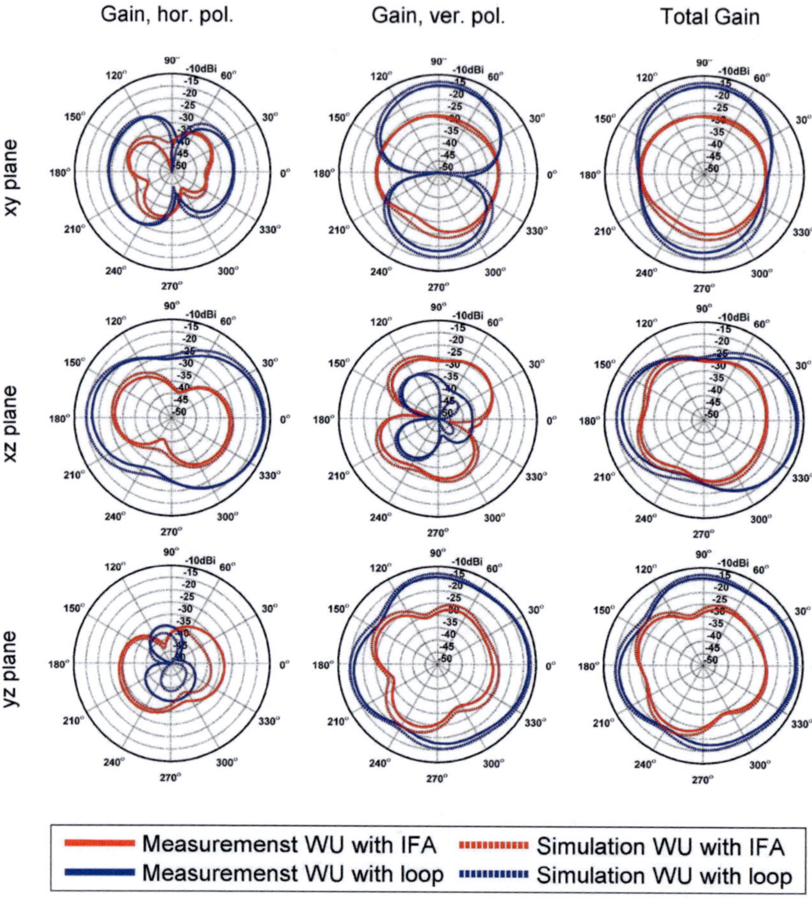

Figure 6.26: Comparison of simulated and measured gain for WU with Inverted F Antenna and WU with loop on the rim (195/65 R15)

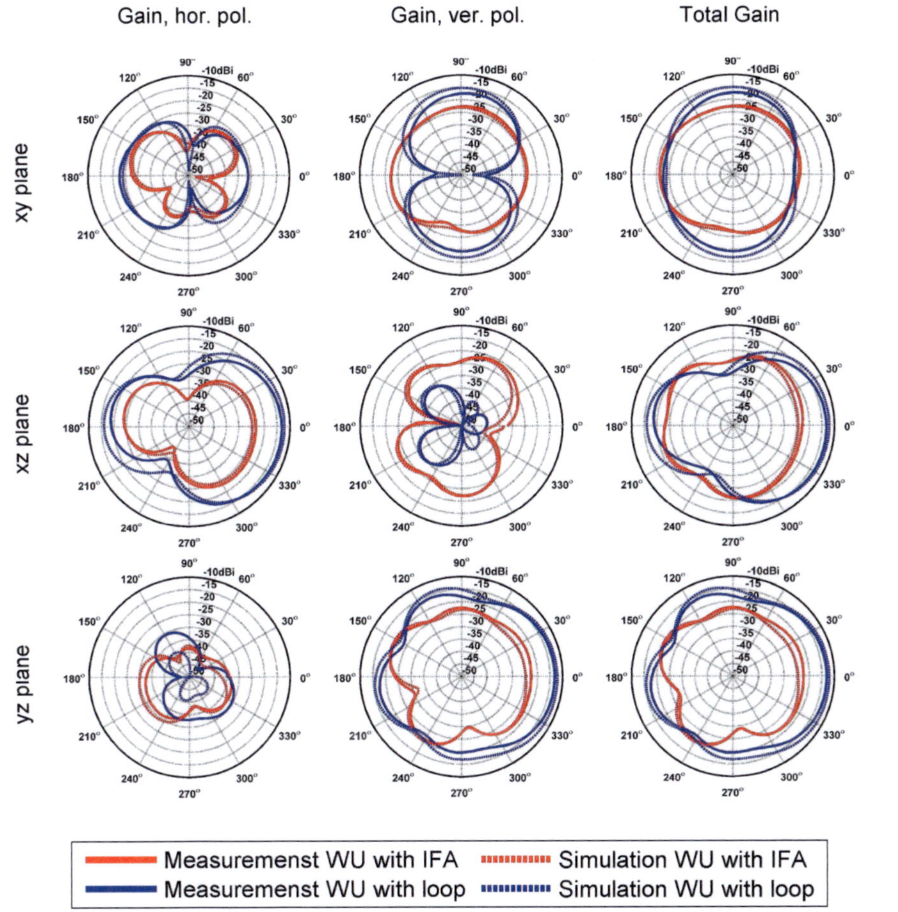

Figure 6.27: Comparison of simulated and measured gain for WU with Inverted F Antenna and WU with loop on the rim (245/40 R18)

The positioning of the loop antenna on the rim is not coincidental. In this case there are two main possible directions to place the WU on the rim. Radiating element (loop) along (Fig. 6.28a) and perpendicular (Fig. 6.28b) to the rim edge. Before the decision was made to place the loop always perpendicular to the rim edge, specific behaviour of this antenna was under investigation. The radiation characteristics improve when the loop is positioned perpendicular to the rim edge. The most significant points of the investigation include following conclusions:

- The loop antenna performance is better on a rim than in free space (stand alone WU).
- Considering radiation characteristics of stand alone WU and on a rim, the antenna Gain increased by 5 dB. This effect is noticeable both on small (R15) and big (R18) rim.
- The loop antenna placed along the rim edge does not deliver as good performance as loop antenna placed perpendicular to the rim edge.

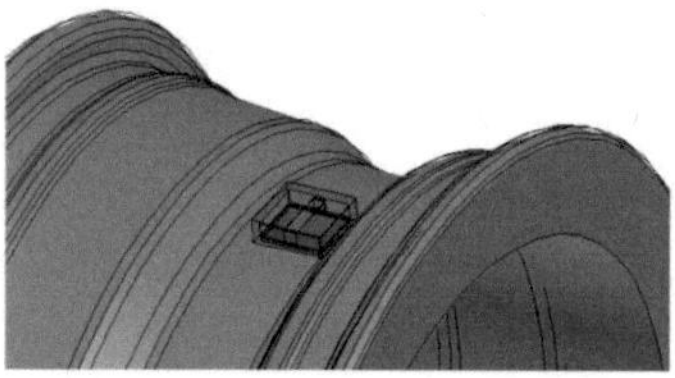

(a) WU with loop along rim edge

(b) WU with loop perpendicular to the rim edge

Figure 6.28: Two possibilities to position loop antenna on a rim

There are few factors that contribute to the above described performance:

- When the WU is placed on a rim, the radiation characteristic becomes more directive. Rim acts as a reflector (well observable in Fig. 6.29).
- Observable change of characteristic antenna parameters, listed in Table 6.5.

Parameter	WU loop stand alone (case 1)	WU loop along the rim edge	WU loop perpendicular to the rim edge (case 2)
G_{max} in dBi	-21.1	-19	-12.1
η in dB	-16.2	-17.3	-9.1
R_{rad} in mΩ	4.5	4.6	23.4

Table 6.5: Characteristic antenna parameters for WU with loop antenna (G_{max} - max. Gain, η - Efficiency, R_{rad} -Radiation Resistance)

The highest achieved gain was in the case of WU with loop antenna placed on a rim in perpendicular orientation to the rim edge. Remarkable is the change in the values of Radiation Resistance. For stand alone WU with loop antenna (case 1) and the same WU placed along the rim edge the Radiation Resistance values are very close with less than 5 mΩ. Radiation Resistance raised to 23.4 mΩ when the WU was placed in perpendicular position to the rim edge (case 2).

Generally the higher the Radiation Resistance, the higher the Radiated Power P_r. There is a linear dependency between radiated power and radiation resistance. Related to a certain antenna point, the Radiation Resistance R_{rad} is equal to the Radiated Power P_{rad} divided by the square of the RMS-value (Root Mean Square) of the antenna current I in this point [88]:

$$R_{rad} = \frac{P_{rad}}{I^2} \tag{6.17}$$

Because of the linear dependency the above mentioned equation can be used to formulate a relationship between two cases: stand alone WU and WU placed perpendicular to the rim edge.

$$I^2 = \frac{P_{rad_{case1}}}{R_{rad_{case1}}} \quad \text{and} \quad I^2 = \frac{P_{rad_{case2}}}{R_{rad_{case2}}} \tag{6.18}$$

$$\Longrightarrow \quad \frac{P_{rad_{case1}}}{P_{rad_{case2}}} = \frac{R_{rad_{case1}}}{R_{rad_{case2}}} \tag{6.19}$$

Herewith follows the difference in the radiated power for both cases:

$$\Delta P_{rad} = 10log\frac{R_{rad_{case1}}}{R_{rad_{case2}}} = -7.16 \text{ dB} \tag{6.20}$$

Comparing the above theoretical calculated value with the difference in the simulated efficiency value for both cases of 7 dB, the outcome of 7.16 dB is very accurate. The value may not be compared directly with the maximum reached Gain for both cases as the radiation characteristic became more directive upon rim influence. The difference in the maximum reached Gain for both cases is higher and amounts 9.1 dB.

- Comparison of WU with loop antenna placed on a rim along the rim edges and perpendicular to the rim edges and their performance is illustrated clearly upon the comparison of the E field distribution in Fig. 6.29. All cases were depicted with the same scale of the field intensity. It is very clear that the magnitude of the E field in the perpendicular case is much higher.

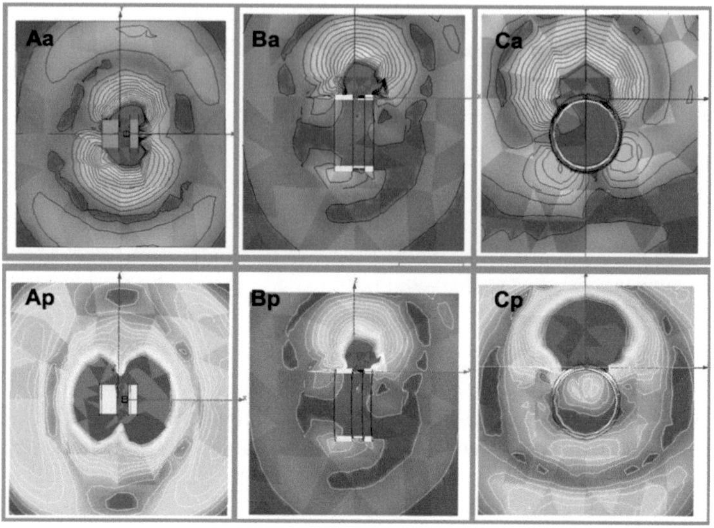

Figure 6.29: Comparison of the E field distribution of a WU with loop antenna on a rim Aa - top view, loop along. Ba - cross section, loop along, Ca - side view, loop along, Ap - top view, loop perpendicular. Bp - cross section, loop perpendicular, Cp - side view, loop perpendicular

Based on the above discussion, it is recommended to place the WU with loop antenna in position loop perpendicular to the rim edge. All further investigations assume such positioning. Investigations of the tyre models and whole TPMS are done in following Chapters with WU with IFA and loop.

6.8.3 WU on the Rim with Conventional Tyres

Good model quality of the entity rim with WUs was assured in the previous Chapter. Next step is the simulation and approval of the prepared models for conventional tyres. In Chapter 3 it was shown that in RF terms there is no difference between conventional summer and winter tyres. Also the tyre brand does not influence the radiation characteristics. Therefore it is required to simulate only one chosen tyre as a representative for all conventional tyres of the chosen size.

Big tyres (245/40 R18) will be presented here as these are further required for full TPMS simulation. As representative tyre Pirelli P Zero Rosso was chosen. Applied electrical parameter values are listed in Table 4.4.

Fig. 6.30 presents measurement as well as simulation results for the chosen tyre with WU IFA in tyre cavity. Fig. 6.31 presents measurement as well as simulation results for the chosen tyre with WU loop antenna in tyre cavity.

After the first analysis of the above presented figures, it may be stated that the conventional tyre models are very accurate. Both the media as well as the skin effect model for the WU with IFA and loop deliver accurate results when compared to the measured values. Model accuracy is confirmed also when analyzing the maximum, minimum and average values listed in Appendix B, Table B.13 for the WU with IFA, Table B.14 for the WU with loop. When compared with stand alone sensors and sensors placed on the equivalent rim (R18), the radiation pattern changed. Both for WU with IFA and loop conventional tyres introduce about 2 to 3 dB losses when compared to the radiation characteristics of these sensors on rim without any tyre.

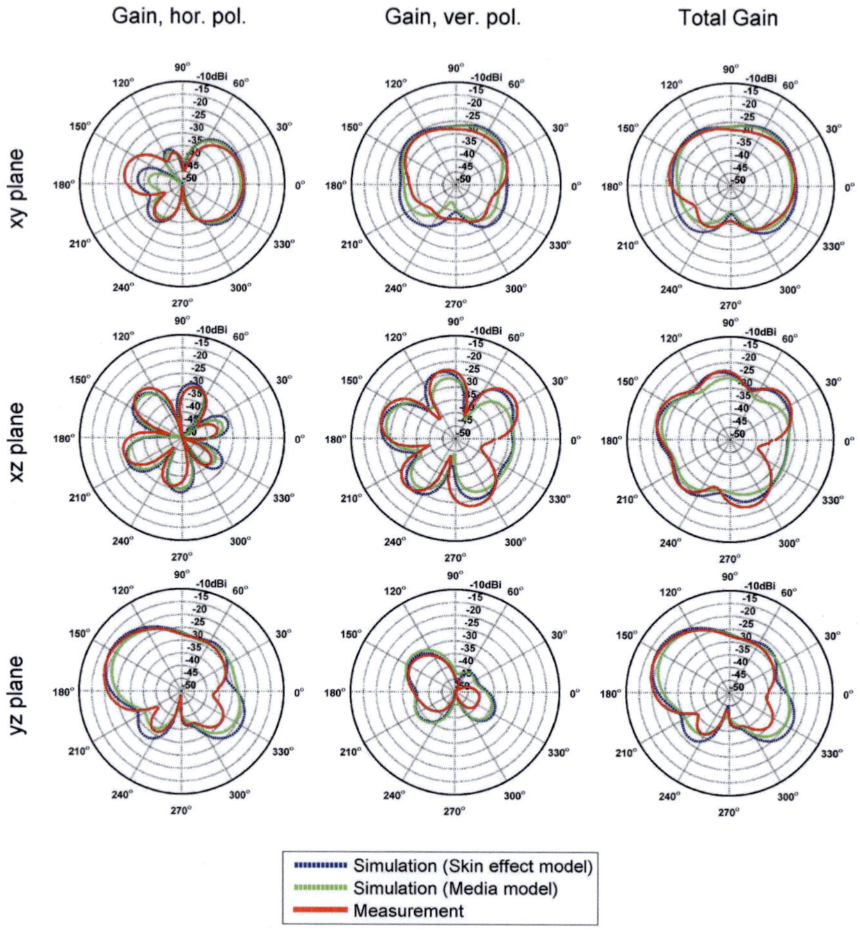

Figure 6.30: Comparison of simulated and measured gain for WU with Inverted F Antenna with conventional representative tyre (big summer tyre 245/40 R18 Pirelli P Zero Rosso)

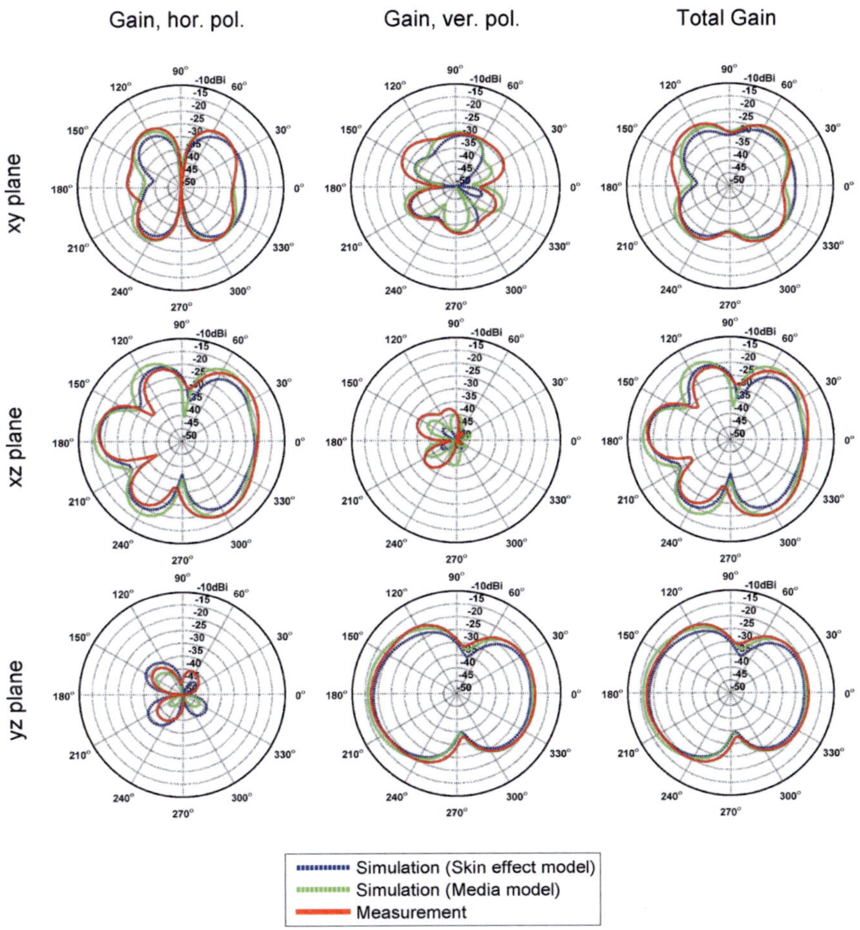

Figure 6.31: Comparison of simulated and measured gain for WU with loop antenna with conventional representative tyre (big summer tyre 245/40 R18 Pirelli P Zero Rosso)

6.8.4 WU on the Rim with Self Supporting Runflat Tyres

Self Supporting Tyre model was presented in Chapter 6.4.3. Simulation and measurement results for representative SSR tyre are depicted in Fig. 6.32 for WU with IFA and in Fig. 6.33 for WU with loop antenna.

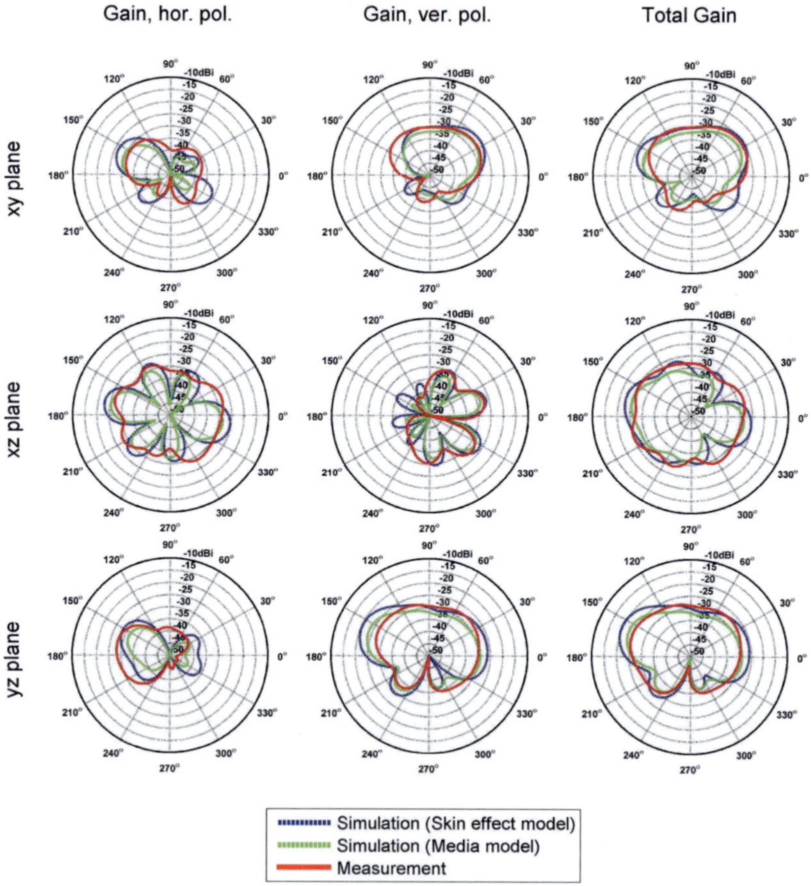

Figure 6.32: Comparison of simulated and measured gain for WU with Inverted F Antenna with SSR representative tyre (big summer tyre 245/40 R18 Bridgestone Potenza REO40)

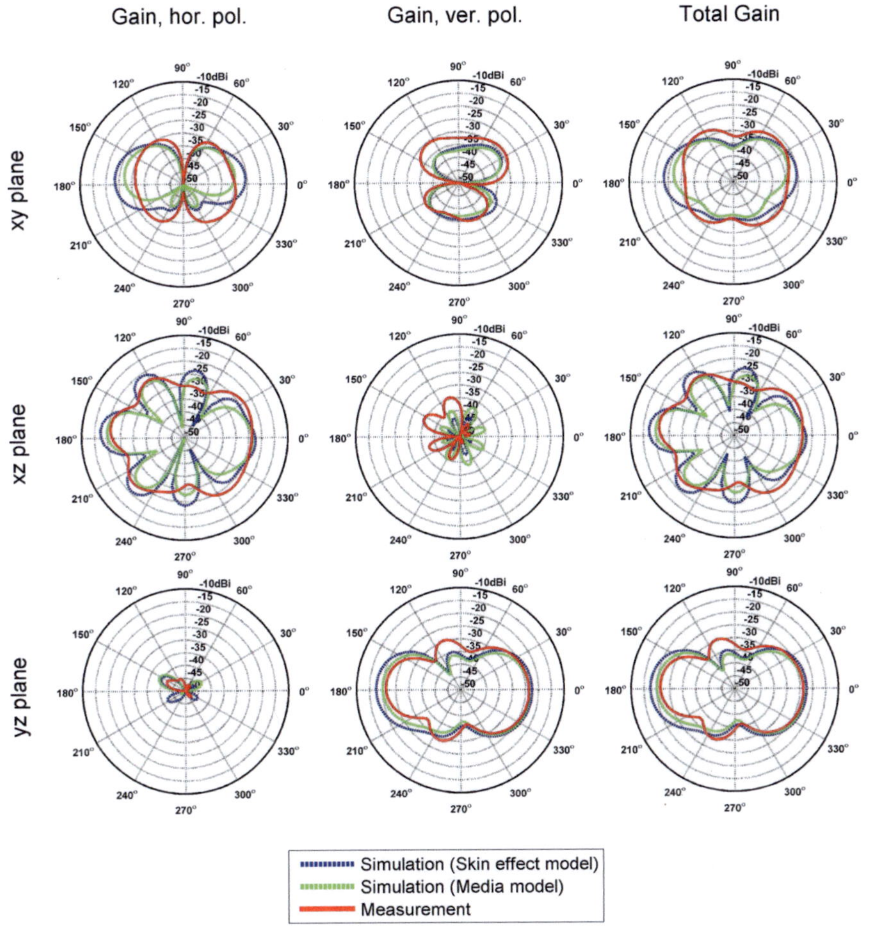

Figure 6.33: Comparison of simulated and measured gain for WU with loop antenna with SSR representative tyre (big summer tyre 245/40 R18 Bridgestone Potenza REO40)

SSR tyre model accuracy in respect to the measurement is good. Maximum, minimum and average values of the achieved Gain may be read from Appendix B, Table B.15 for WU with IFA and Table B.16 for loop antenna. The characteristic for of the radiation pattern changed in comparison to the conventional tyres. SSR tyres feature more attenuation effect than the conventional tyres, on average both in case of WU with IFA and with loop the amplitude dropped by approx. 5dB.

6.9 TPMS System Characterization

Up to this point, all TPMS system components were simulated and the accuracy of the models was approved for single components such as WU, WU on the rim with different tyre types. also receiver performance was approved. The vehicle body and the interaction of these components as well as transmission path was not examined yet.

This Chapter deals with the whole TPMS system performance test including measurement repeatability, TPMS performance dependency on the tyre type, correlation between the simulated and measured results.

6.9.1 Measurement Setup

TPMS measurements were conducted in a controlled environment on a dedicated test site. For test repeatability the vehicle was pulled at a constant speed with the help of a winch. Driver or passenger - generally human influence was not taken into account. Equipment that composed the measurement setup is presented in Fig. 6.34.

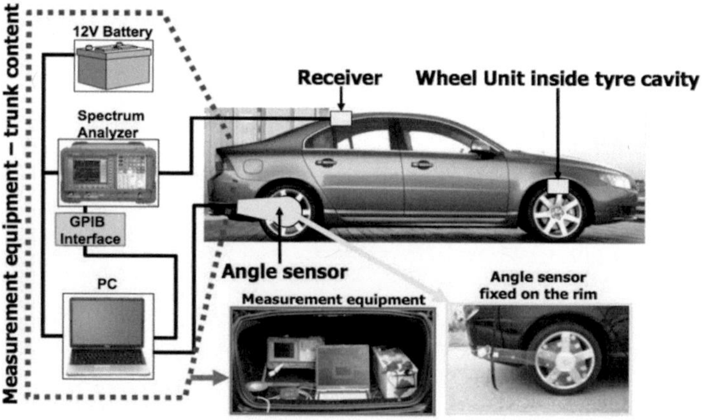

Figure 6.34: TPMS measurement setup

The vehicle, WU and the receiver belong to the system under test. All measurements were done with the same vehicle as well as the same receiver. Different tyres and WUs were in use. The influence of different vehicle bodies is not a subject of this work.

In order to minimize the influence of the measurement equipment on the results it was positioned in the trunk. The phenomenon of Faraday cage [127] was used in order to minimize the interference that could occur. Coaxial cable from the receiver was laid following the metallic vehicle structure along the roof and C-pillar. Measured attenuation of the applied coaxial cable of 1.2 dB was taken into account during system evaluation.

Measurement equipment includes the spectrum analyser connected via GPIB interface with the PC (LabView virtual instrument which gathers the data from the spectrum analyzer and the angle sensor) and power supply. The spectrum analyzer measured the varying received power in dependency on the wheel position. In order to know the exact wheel position and therewith the WU position an angle sensor was applied. The measurements were conducted for each of the four wheels separately.

6.9.2 Measurement Repeatability

Every TPMS measurement was started with the same wheel position and the received power was measured over three full wheel turns to check and assure result repeatability. The measurements started with the WU positioned at zero degrees indicated in Fig. 6.35a.

(a) Wheel rotation direction and sensor position indication

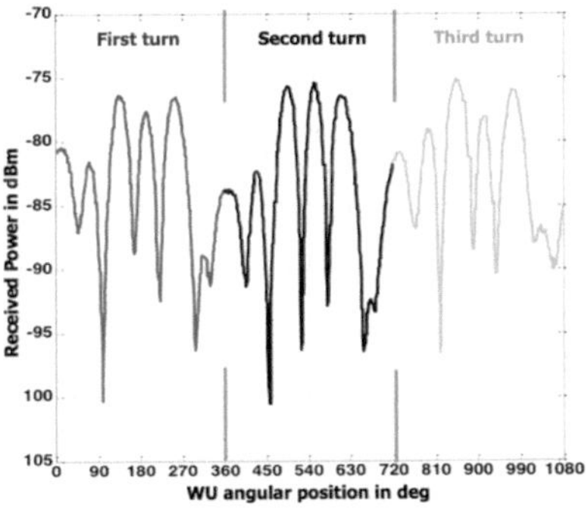

(b) Example measurement result - received power from WU in conventional tyre cavity over three wheel rotations

Figure 6.35: Wheel rotation

Example measurement of the received WU power in conventional tyre cavity (245/40 R18) is presented in Fig. 6.35b. The three consequent wheel turns deliver slightly different results when the amplitude of the received power is considered. The difference between each wheel turn is in the range of 3 dB. For all further results an average over three consequent wheel turns is calculated.

6.9.3 TPMS Characterization with Conventional Tyres

TPMS system was characterized with conventional tyre type for: the WU with IFA and loop antenna. The measured profile of the radio frequency channel of propagation is presented in Fig. 6.36 for WU with IFA and in Fig. 6.37 for WU with loop.

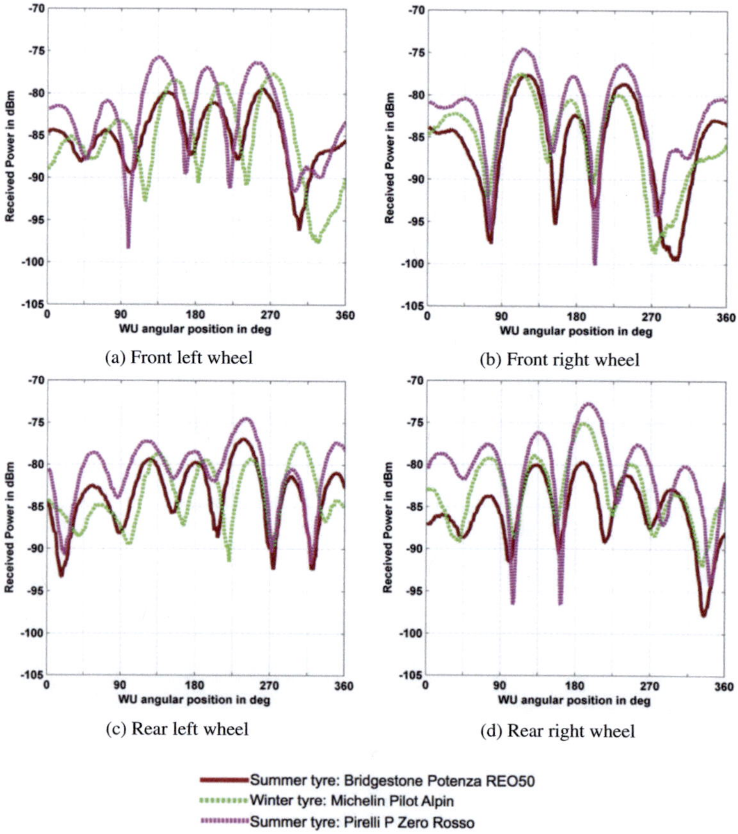

Figure 6.36: Measured received power over the angular sensor position for conventional tyre 245/40 R18 with WU IFA in tyre cavity

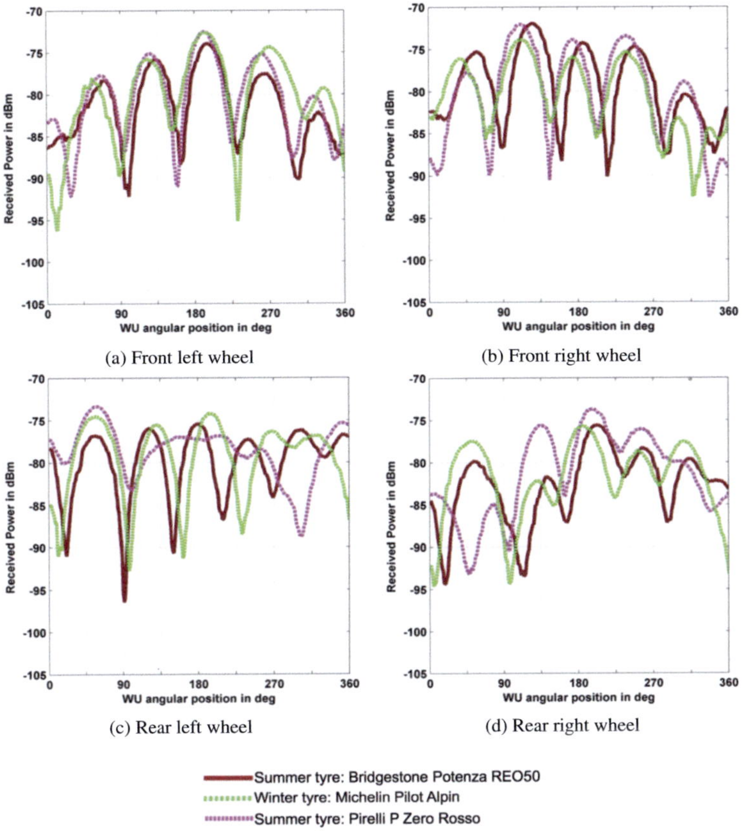

Figure 6.37: Measured received power over the angular sensor position for conventional tyre 245/40 R18 with WU loop in tyre cavity

TPMS system was characterized for both WU sensors with IFA and loop with the help of the same example tyres. Received power varies in dependency on the angular WU position. Following conclusions and knowledge was obtained:

- All tyres of one construction type (here presented conventional tyres) deliver similar characteristics of the RF TPMS channel.
- Each wheel features different channel characteristic. Received power level has to be characterized for each wheel separately.

- WU equipped with loop antenna delivers higher Received Power than the WU equipped with IFA. Power drops for the WU with loop are not so deep as for WU with IFA.

- In both WU cases generally the amplitude of the Received Power is higher for WU positions between 90 and 270 deg. This means when the WU leaves the wheelhouse and is closer to the ground.

6.9.4 TPMS Full System Simulation

TPMS full system simulation requires high computational resources. Therefore in scope of this work full system simulation was conducted only for one wheel. Simulation result for left front wheel with conventional tyre and WU with IFA is depicted in Fig. 6.38a. Analytical model for TPMS full system simulation is depicted in Fig. 6.38b.

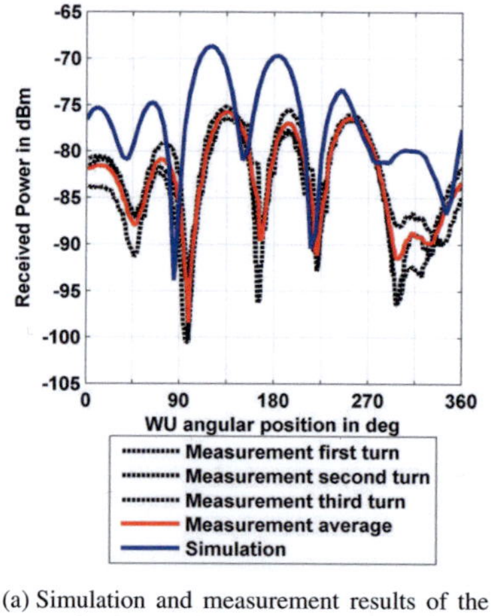

(a) Simulation and measurement results of the received power between the WU with IFA in left front conventional tyre cavity and the receiver

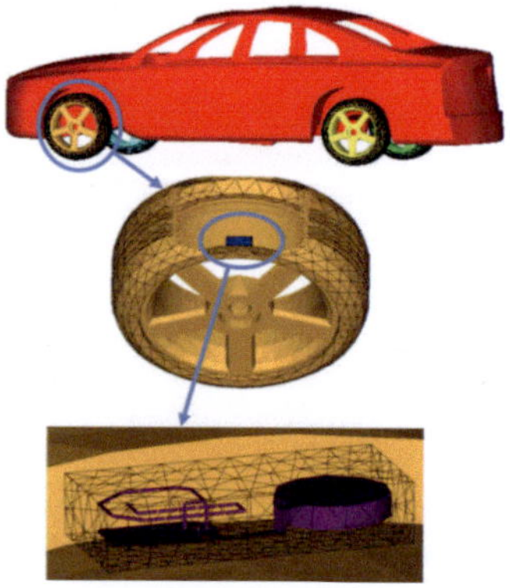

(b) Analytical model for TPMS computation

Figure 6.38: TPMS simulation

It may be observed that also full TPMS simulation confirms high accuracy of the prepared parametric models. Simulated received power curve progression follows very well the measured RF channel characteristic. Amplitude offset of approx. 6 dB is caused by the fact that the vehicle interior influence is not considered in the simulation process.

Essential is the existence of the RF link between the receiver (vehicle) and WU placed in tyre cavity. With the above achieved results main factors that have impact on the TPMS RF propagation are known. It is clear in which way do such elements as WU, rim, tyre and vehicle body influence the system performance. It is possible to formulate the Link Budget for TPMS with following formula:

$$S_r \geq P_t + G_{WU} + A_{RIM} + A_{TYRE} + A_{VEHICLE} + F_s \tag{6.21}$$

The receiver system sensitivity S_r has to be in a certain balance, at least equal or in the optimal case greater than following listed factors.

- The transmitted power at the feeding point of the WU antenna P_t.
- The average gain of the WU antenna G_{WU}.
- Rim influence A_{RIM}.
- Tyre influence A_{TYRE}.
- Vehicle body influence $A_{VEHICLE}$.
- The reliability system factor F_s, which comprises tolerance of all system components and ensures confidence margin for the system functionality as well as human influence.

Additionally system performance is dependent on the vehicle environment and it should be always assured by appropriate measurements in different conditions. This influence should be taken into account when setting the level of system reliability factor F_s upon experience.

Based on the considerations of all factors analyzed in this thesis following example values and limits presented in Table 6.6 may be considered.

The value of transmitted power P_t is system specific and very often regulated by national law of each country. Therefore this value may not be treated as a generic one and will be specific for each new designed system. In order to ease the comparison between the analyzed WUs it was set to the same value of 5.5 dBm.

Achieved average Gain of a single stand alone WU is specific for the antenna design and may be approved with simulation, measurement or theoretical calculation. The here achieved average values of G_{WU} are for IFA -16.6 dBi (fulfilled the requirements) and for loop -23.1 dBi (approved for further investigations due to antenna type (magnetic), see Chapter 5.3). These values may also not be treated as generic ones but design specific.

<table>
<tr><th>Antenna</th><th>P_t</th><th>$avg.G_{WU}$</th><th>A_{RIM}</th><th>A_{TYRE}</th><th>$A_{VEH.}$</th></tr>
<tr><td rowspan="2">IFA</td><td rowspan="2">5.5 dBm</td><td rowspan="2">-16.6 dBi</td><td rowspan="2">-6.1 dB</td><td>Conv.: -7.7 dB</td><td rowspan="2">-10 dB</td></tr>
<tr><td>SSR: -11 dB</td></tr>
<tr><td rowspan="2">LOOP</td><td rowspan="2">5.5 dBm</td><td rowspan="2">-23.1 dBi</td><td rowspan="2">5.1 dB</td><td>Conv.: -3.7 dB</td><td rowspan="2">-10 dB</td></tr>
<tr><td>SSR: -8.5 dB</td></tr>
</table>

Table 6.6: Characteristic values of characterized TPMS system

Rim influence is described by A_{RIM}. Within the underlying work two rim sizes were under investigation. For electric antenna types (e.g. IFA) rim introduces losses, whereas for magnetic antenna type (e.g. loop) introduces amplification. This phenomenon was observable for both rim sizes. Analyzed WU with IFA antenna lost 6.1 dB in Gain on average and WU with loop antenna gained 5.1 dB. The rim influence may be treated as general true for the in the future designed systems.

Tyre influence is described with A_{TYRE}. Three tyre types were analyzed: slick, conventional and Self Supporting Runflat. Firstly slick tyre model was approved as the tyre with the least complicated structure. This tyre type is however only used for special applications as car racing. For this reason only one single tyre was analyzed and allowed investigation of the accuracy for more complicated tyre models. Generally tyres influence much more the electric antenna types as the magnetic ones. For WU with IFA conventional tyres introduce approx. 8 dB and SSR tyres 11 dB loss. In case of WU with loop antenna these are approx. 4 dB and 9 dB equivalent for conventional and SSR tyres.

Vehicle influence $A_{VEHICLE}$ in the specific analyzed case introduced the loss of 10 dB due to vehicle interior parts.

The here described values for the Link Budget present a guideline and basis for future system development and analysis.

7 Conclusions

The here presented work contributes to the advancement in RF automotive examination. Application of low / high-end wired and wireless electronics emerges meanwhile almost all in vehicle integrated functions. Within the scope of this work examined were such systems as PASE, RKE and TPMS. All considerations were exemplified on one predefined vehicle body.

Important to stress is that all system components and their influence on the overall system performance were considered in the work. This includes vehicle specific parts as chassis, rim and tyres, as well as system specific components as receiver module and WUs. Examination of the described systems functionality incorporated both practical measurements and simulative tasks.

The main focus was put not only on the system performance improvement and quality assurance, but mostly on the reduction of the time dedicated to the proceeding development steps. These steps usually require many iterative measurements both in the preconditioned laboratory environment as well as vehicle application tests. The main challenge was to create a stable simulation environment for all systems incorporating library with all vehicle and system parts. Standard, parametric simulation models were created that allow meaningful reduction of preparation and simulation time. In order to achieve this, it was essential to understand the influence of all system components and of the applied materials (dielectric material characterization) on RF overall system performance.

Reflecting shortly to PASE, RKE and TPMS systems in detail, Fig. 7.1 presents briefly all system components under consideration and the most important conclusions.

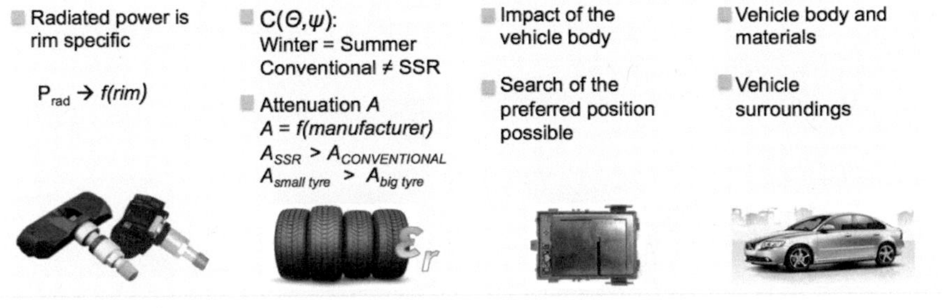

Figure 7.1: Result overview and analyzed system component presentation

Upon the study following conclusions reveal:

- The radiated power from a Wheel Unit is dependent on the rim size and form.
- There is no difference in the radiated characteristic independent whether winter or summer tyres are in use.
- Conventional and SSR tyres feature different radiation characteristic.
- Tyre attenuation is specific for each tyre brand.
- SSR tyres contribute to more attenuation than conventional tyres.
- Small tyres reveal more attenuation than bigger tyres.
- Vehicle body and the surrounding materials influence the performance of the receiver module.
- An algorithm was presented how it is possible to search for a preferred, optimum receiver position.
- Vehicle structure and vehicle surrounding influence system performance.

With the presented thesis RF simulations became an integral part in the system design and the application process. Following advantages were achieved with the help of the underlying work:

- Better overall system understanding.
- Prediction of system behaviour in all steps of product lifecycle: predevelopment, development, application, SOP and maintenance.
- RF system parameter control (link budget, range, module (antenna) position, antenna gain, attenuation (material influence)).
- Easy prediction of system performance if any changes occur (new vehicle body, facelift, harness guidance, new tyre models, new rim models, packaging variations, new applied materials etc.).
- Simplified and fast evaluation of automotive RF components thanks to pre-configured simulation models.
- Avoidance of costly iterative processes and mistakes.
- Controlled system reliability.
- Reduction of development and application time.

The here presented work for the first time treats all three systems with their complexity and confirms possibility to deliver parametric, reusable RF simulation models. The aim to reduce effort significantly for the system design and application was achieved.

Here described simulation methodology may be applied to any other RF automotive system and the achieved specific results for PASE, RKE and TPMS are to be treated as a showcase incorporating different frequency ranges and vehicle body elements (varying range of the ratio element size and wavelength). All aspects of complex automotive simulation processes (element classification, material influence and characterization, model simplification, radiation characteristics, etc.) were taken into account and form a guide for automotive engineering industry. The work presents the "how to" electromagnetic automotive simulation and forms the basis, for the system design and optimization for vehicle manufacturers as well as system suppliers.

Bibliography

[1] *"Automotive Technology: Greener Vehicles, Changing Skills"*, Electronics, Software and Controls Report, CAR Center for Automotive Research, 2011

[2] M. Brzeska, G. -A. Chakam *"Compact Dual-band Reduced Size PIFA Antenna for Automotive Applications"*, Proceedings of the European Conference on Antennas and Propagation (EuCAP), 2006

[3] M. Brzeska, J Pontes, G. -A. Chakam, W. Wiesbeck, *"Modellierung und messtechnische Charakterisierung von Reifendrucksensoren"*, Sensoren im Automobil II, Haus der Technik Fachbuch Band 86, Expert Verlag, 2007

[4] M. Brzeska, G. -A. Chakam *"Modeling of the coverage range for modern vehicle access systems at low frequencies"*, Proceedings of the 37th European Microwave Conference, EuMA, 2007

[5] S. Sevskiy, M. Brzeska, W. Wiesbeck *"Multi-band dipole array for WLAN access points"*, Technical Report for Joint COST 273/284, Gothenburg, 2004

[6] M. Brzeska, S. Sevskiy, W. Wiesbeck *"Ultra-wideband and multi-band antennas for wireless communications"*, Proceedings of the International Astronautical Congress, Vancouver, Canada, 2004

[7] M. Brzeska, G. -A. Chakam *"Measurement, simulation and analytical Modeling of the Radiowave Propagation Environment for the Automobile Access Systems"*, Proceedings of the European Conference in Antennas and Propagation (EuCAP), Niece, France, 2006

[8] G. -A. Chakam, M. Brzeska, A. Schaefer, M. Weinberger, A. Arango *"Comparison of key-fob antennas for vehicle access system"*, Proceedings of the European Conference in Antennas and Propagation (EuCAP), Niece, France, 2007

[9] M. Brzeska, J Pontes, G. -A. Chakam, W. Wiesbeck, *"RF-Design, characterization and modeling of tyre pressure sensors"*, Proceedings of the European Conference on Antennas and Propagation, Edinburgh, United Kingdom, 2007

[10] M. Brzeska, G. -A. Chakam *"RF-Modeling and characterization of a tyre pressure monitoring system"*, Proceedings of the European Conference on Antennas and Propagation, Edinburgh, United Kingdom, 2007

[11] G. -A. Chakam, M. Brzeska *"Achieving Long Range Communication in Cars"*, Proceedings of the European Conference on Antennas and Propagation, Edinburgh, United Kingdom, 2007

[12] M. Brzeska, R. Quintero, V. Mata, C. Domingo-Sabat, E. Martinez *"Novel Fractal Solution for Integrated TMC Antenna"*, Proceedings of the 3rd European Conference on Antennas and Propagation, Berlin, Germany, 2009

[13] M. Brzeska, R. Quintero, E. Martinez, L. Tantina, R. Cisneros, D. Puy *"New Generation of In-Mirror Integrated Antennas"*, Proceedings of the 3rd European Conference on Antennas and Propagation, Berlin, Germany, 2009

[14] M. Thumm, W. Wiesbeck, S. Kern *"Hochfrequenzmesstechnik: Verfahren und Messsysteme"*, Teubner Verlag, 1998

[15] J. Grosinger, L.W. Mayer, C.F. Mecklenbraeuker, A.L. Scholtz *"Input impedance measurement of a dipole antenna mounted on a car tire"*, International Symposium on Antennas and Propagation, 2009

[16] H. Zeng, T. Hubing *"Investigation of Loop and Whip Antennas in Tire Pressure Monitoring Systems"*, Vehicular Technology Conference, 2012

[17] US National Highway Traffic Safety Administration, *"Traffic Safety Facts, Research Note"*, US. Department of Transportation, National Highway Traffic Safety Administration, 2012

[18] Organisation Internationale des Constructeurs d'Automobile, International Organization of Motor Vehicle Manufacturers, Production Statistics, 1997 - 2010

[19] R. Stein, *"The Automobile Book"*, Paul Hamlyn Ltd., 1967

[20] M. Ceretelli, G. Biffi Gentili, *"Progress in Compact Multifunction Automotive Antennas"*, IEEE, 2007

[21] Freescale Semiconductor *"Freescale Single Package Tyre Pressure Monitoring System (TPMS)"*, Document Number TPMSWP, 2007

[22] M. Beecham, *"Vehicle access and security system trends, forecasts and companies"*, 2005

[23] Continental Automotive GmbH, *"Interior, Daten und Fakten 2009"*, 2009

[24] Maxim, *"Requirements of Remote Keyless Entry (RKE) Systems, APPLICATION NOTE 3395"*, Maxim Integrated Products, 2008

[25] J. L. Volakis *"Antenna engineering handbook"*, The McGraw-Hill Companies, 2007

[26] National Highway Traffic Safety Administration, *"Tyre Pressure Monitorin Final Rule"*, URL: http://www.nhtsa.dot.gov, 2009

[27] C. Burrer *"Direct measuring tire pressure monitoring systems"*, VDI Berichte, pp. 517-521, 2004

[28] V. Kukshya, H. J. Song, H. P. Hsu, R. W. Wiese *"Characterizing Performance of Tire Pressure Monitoring Systems Using Experimental Measurements and System simulations"*, IEEE, 2007

[29] M. Fischer *"Tire Pressure Monitoring, Design and functionality of direct measuring systems"*, Die Bibliothek der Technik Vol. 243, 2003

[30] E. Hirose, N. Michishita, Y. Yamada, *"Radiated Waves from a Transmitter contained in a Vehicle Tire"*, Antennas and Propagation Society International Symposium, IEEE, 2006

[31] M. Hanlon *"EarthFirst - promising new tyre recycling technology"*, 2006

[32] *"Tires and Passenger Vehicle Fuel Economy: Informing Consumers, Improving Performance - Special Report 286"*, 2007

[33] *"Improved Goodyear Run-flat Tires Deliver Lexus Luxury"*, URL: http://www.AutoChannel.com, 2001

[34] *"Amerityre Issues Update on FMVSS Testing for Run-Flat Tires"*, 2007

[35] Continental Automotive GmbH, URL: http://www.continental.de, 2007

[36] F.R. Eirich *"Science and technology of rubber"*, Rubber division of the American Chemical Society, Academi Press, 1994

[37] D. Dimitrova *"Modellierung, Simulation und Analyse der Bestandteile eines Reifendruckkontrollsystems"*, Diplomarbeit, Fachhochschule Deggendorf, 2007

[38] Global Market Review *"A review of the European tyre market - Management briefing"*, 2004

[39] C.A. Balanis *"Antenna Theory, Analysis and Design"*, John Wiley & Sons, Inc., ISBN 0-471-59268-4, 2nd Ed. 1982

[40] D. K. Cheng *"Field and Wave Electromagnetics"*, Addison-Wesley Publishing Company Inc., ISBN 0-201-52820-7, Edition 2, 1998

[41] E. C. Jordan, K.G. Balmain *"Electromagnetic Waves and Radiating Systems"*, Prentice-Hall, ISBN 81-203-0054-8, 1968

[42] H. A. Wheeler, *"The radiansphere around a small antenna"*, Proc. IRE, S. 1325-1331, Aug. 1959

[43] J. A. Shaw *"Radiometry and the Friis Transmission Equation"*, IEEE, 2005

[44] S. I.Ganchev, J. Bhattacharyya, S. Bakhtiari, N. Qaddoumi, D. Brandenburg, R. Zoughi *"Microwave Diagnosis of Rubber Compounds"*, IEEE Tansactions on Microwave Theory and Techniques, Vol. 42, No. 1, 1994

[45] V. Komarov, S. Wang, J. Tang *"Permittivity and Measurement"*, Washington State University

[46] K. Hirasawa, M. Haneishi *"Analysis, Design, and Measurement of Small and Low-Profile Antennas"*, Artech House, 1992

[47] G. W. Hippisley, J. Carr *"Practical Antenna Handbook"*, McGraw-Hill, 2010

[48] R. D. Straw *"The ARRL antenna book: The Ultimate Reference for Amateur Radio Antennas"*, American Radio Relay League, ARRL, 2003

[49] G. Breed, *"Basic Principles of Electrically Small Antennas"*, High Frequency Electronics, Summit Technical Media, LCC, 2007

[50] D. Miron, *"Small Antenna Design"*, Newnes, 2006

[51] A. Skrivervik, J.-F. Zuercher, O. Staub, J.R. Mosig *"PCS Antenna Design: The Challenge of Miniaturization"*, IEEE Antennas and Propagation Magazine, Vol. 43, No. 4, 2001

[52] J. P. Gianvittorio, Y. Rahmat-Samii *"Fractal Antennas: A Novel Antenna Miniaturization Technique and Applications"*, IEEE Antennas and Propagation Magazine, Vol. 44, No. 1, 2002

[53] R. Waterhouse *"Printed antennas for wireless communications"*, John Wiley & Sons Ltd., 2007

[54] Y. Yasunaga, F. Watanabe, T. Shiokawa, H. Yanada *"Phased Array Antennas for Aeronautical Satellite Communications"*, IEE, ICAP 87, 1987

[55] R. F. Harrington *"Effect of Antenna Size on Gain, Bandwidth, and Efficiency"*, Journal of Research of the National Bureau of Standards-D, Radio Propagation, Vol. 64D, No. 1, January-February 1960

[56] G.-A. Chakam *"An Inductively Loaded PIFA for Automotive Applications"*, European Microwave Week, Amsterdam, 2004

[57] Z.D. Liu, P.S. Hall, D. Wake *"Dual-frequency planar inverted-F antenna"*, IEEE Transactions Antennas Propagation, 1997

[58] K. Hirasawa, M. Haneishi *"Analysis, Design and Measurement of small and Low-profile Antennas"*, Artech House, UK, 1992

[59] R. Leelaratne, R. Langley *"Multiband PIFA vehicle telematics antennas"*, IEEE Transactions on Vehicular Technology, 2005

[60] T. Taga, K. Tsunekwa *"Performance analysis of a built-in planar inverted f antenna for 800 MHz band portable radio units"*, IEEE Transactions, 1987

[61] K. L. Vigra, Y. Rahmat-Samii *"Low-profile enhanced bandwidth PIFA antennas for wireless communications packaging "*, IEEE Transactions, 1997

[62] M.A. Jensen, Y. Rahmat-Samii, *"FDTD analysis of PIFA diversity antennas on a hand held transceiver unit"*, IEEE Antenna Propagation Symposium, Atlanta, 1993

[63] K. L. Wong, *"Compact and broadband microstrip antennas"*, John Wiley & Sons, Inc., New York, 2002

[64] L. Calhau1, P. Pinho, *"Low Profile Multi-Band Antenna for Mobile Communications"*, Proceedings European Conference on Antennas and Propagation (EuCAP), 2010

[65] J.S. Kuo, K.L. Wong, *"Dual frequency operation of planar inverted L antenna with tapered patch width"*, Microwave Opt Technol Lett, 1999.

[66] F. Yang, X.X. Zhang, X. Ye, Y. Rahmat-Samii, *"Wide-band E-shaped patch antennas for wireless communications"*, IEEE Transactions Antennas Propagation, Vol. 49, pp. 1094 minus 1100, 2001

[67] Continental AG URL: http://www.conti-online.com, 2011

[68] Y. Huang, K. Boyle, *"Antennas: From Theory to Practice"*, John Wiley and Sons Ltd., 2008

[69] S.R. Saunders, A.A. Zavala, *"Antennas and Propagation for Wireless Communication Systems"*, John Wiley and Sons Ltd., 2007

[70] B. Allen, M. Dohler, E. Okon, W. Malik, A. Brown, D. Edwards, *"Ultra Wideband Antennas and Propagation for Communications, Radar and Imaging"*, John Wiley and Sons Ltd., 2006

[71] *"UHF ASK/FSK transmitter ATA5756/5656 manual"*, Atmel, Rev. 4702H-RKE, 2004

[72] S. Jawale, *"RF Modelling and Characeterizing of a Tyre Guard Sensor used for Tyre Pressure Monitoring"*, Diplomarbeit, Universität Karlsruhe (TH), Institut für Höchstfrequenztechnik und Elektronik, 2005

[73] C. Icheln, P. Vainikainen, P. Haapala, *"Application of GTEM cell to small antenna measurements"*, IEEE Antenna and Propagation Society International Symposium Digest, Vol. 1, 1997

[74] S.W. Su, K.L. Wong, Y.T. Cheng, W.S. Chen, *"Finite-ground-plane effects on the ultra-wideband planar monopole antenna"*, Microwave and Optical Technology Letters, 2004

[75] M.A. Jensen, Y. Rahmat-Samii, *"Performance analysis of antennas for hand-held transceivers using FDTD"*, IEEE Transactions on Antennas and Propagation, Vol. 42, 1994

[76] K. Fujimoto, A. Henderson, K. Hirasawa, J.R. James, *"Small Antennas"*, U.K. Research Studies, 1987

[77] J.A. Flint, *"Numerical Analysis of Detuning in Printed Inverted F Antennas for Bluetooth Applications"*, European Personal Mobile Communications Conference, 2003

[78] S. Cloude, *"An Introduction to Electromagnetic Wave Propagation and Antennas"*, UCL Press Limited, 1995

[79] K.A. Bakshi, A.V. Bakshi, U.A. Bakshi *"Antennas And Wave Propagation"*, Technical Publications Pune, 2008

[80] M. Olmos, H.D. Hristov, R. Feick, *"Inverted-F antennas with wideband match Performance"*, Electronic Letters, Vol. 38, 2002

[81] Ethertronics internal report 10019A *"Antenna loss inside enclosures"*, 2000

[82] M. Golio *"The RF and microwave handbook"*, CRC Press, 2001

[83] G.S. Smith *"An Introduction to Classical Electromagnetic Radiation"*, Cambridge University Press, 1997

[84] H. Sizun *"Radio wave propagation for telecommunication applications"*, Springer Verlag, 2003

[85] I. Walter, E. Arnold, C. Holtzhaussen, M. Sabielny *"Antenna Coupling on Electromagnetic Large Objects"*, EADS Germany, GeMiC 2005

[86] M. Nisznansky *"Modelling and Optimization of Broadband Electrical Interconnects and Housings for Optical Communication Systems"*, Faculty of Engineering Sciences of Friedrich-Alexander University Erlangen-Nuremberg, 2005

[87] S.J. Orfanidis *"Electromagnetic Waves and Antennas"*, ECE Department, Rutgers University, URL: http://www.ece.rutgers.edu

[88] K. Rothammel, A. Kirschke *"Antennen Buch"*, DARC Verlag, 2002

[89] W. Storr *"Electronics Tutorials"*, URL: http://www.electronics-tutorials.ws/diode/diode_1.html, 2010

[90] H.S. Nalwa *"Handbook of Advanced Electronic and Photonic Materials and Devices"*, Academic Press, 2001

[91] D.S. Nyce *"Linear Position Sensors, Theory and Application"*, John Wiley and Sons Inc., 2004

[92] J. Fraden *"Handbook of Modern Sensors, Physics, Design and Applications"*, Advanced Monitors Corporation, 2003

[93] M.J. Loboda, R. Singh, S.S. Ang, H.S. Rathore *"Low and High Dielectric Constant Materials: Materials Science, Processing and Reliability Issues"*, The Electrochemical Society, 2000

[94] D. Fleisch *"A Student's Guide to Maxwell Equations"*, Cambridge University Press, 2008

[95] L.F. Chen, C.K. Ong, C.P. Neo, V.V. Varadan, V.K. Varadan *"Microwave electronics: measurement and materials characterization"*, John Wiley and Sons Inc., 2004

[96] S. Begley, *"Electromagnetic Properties of Materials: Characterization at Microwave Frequencies and Beyond"*, Application Development Engineer, Agilent Technologies, 2010

[97] C. Beadle *"Rubber; Production and Utilisation of the Raw Product"*, Jesson Press, 2008

[98] M.S. Venkatesh, G.S.V. Raghavan *"An overview of dielectric properties measuring techniques"*, Canadian Biosystems Engineering, Vol. 47, 2005

[99] J. Baker Jarvis, M.D. Janezic, B.F. Riddle, R.T. Johnk, P. Kabos, C. Holloway, R.G. Geyer, C.A.Grosvenor, *"Measurement of the Permittivity and Permeability of Lossy Materials: Solids, Liquids, Metals, Building Materials and Negative-Index Materials"*, National Institute of Standards and Technology (NIST), Technical Note 1536, Boulder, 2004

[100] T. Sphicopoulos, V. Teodoridis, F.E. Gardiol, *"Simple Nondestructive Method for the Measurement of the Material Permittivity"*, Journal for Microwave Power, 1985

[101] B.L.J. Kulesza, J.S. Thorp, A. Bakar Ahmad *"Coaxial line methods for measuring permittivity and dielectric loss"*, Journal of Materials Science, Vol. 19, No. 3, 1984

[102] R. Zajicek, J. Vrba, K. Novotny *"Evaluation of a Reflection Method on an Open-Ended Coaxial Line and its Use in Dielectric Measurements"*, Acta Polytechnica Vol. 46, No. 5, Czech Technical University Publishing House, 2006

[103] *"Intelligent Tyre for Accident-free Traffic, Intelligent Tyre Systems - State of the Art and Potential Technologies"*, Apollo, Deliverable D7, 2001

[104] L. Ibbotson *"The Fundamentals of Signal Transmission"*, Elsevier Ltd, 1999

[105] J.C. Whitaker *"The Electronics Handbook"*, CRC Press, 2005

[106] R.F. Graf *"Modern Dictionary of Electronics"*, Newnes, 1997

[107] V.A. Bakshi, A.V. Bakshi *"Transmission Lines And Waveguide"*, Technical Publications Pune, 2009

[108] L. Ganesan, S. Mole *"Transmission Lines & Waveguide"*, Tata McGrow Hill, 2010

[109] T. Sabu, J.S Kuruvilla. K. Malhotra, G. Koichi, M.S. Sreekala *"Polymer Composites, Macro- and Microcomposites"*, Wiley, 2012

[110] B. Bhushan *"Handbook of Nanotechnology"*, Springer, 2007

[111] H. Czichos, T. Saito, E. Smith *"Handbook of Metrology and Testing"*, Springer, 2011

[112] D.A. Sanchez-Hernandez *"High Frequency Electromagnetic Dosimetry"*, Artech House Inc., 2009

[113] T.S. Mailadil *"Dielectric Materials for Wireless Communication"*, Elsevier, 2008

[114] S.A. Dyer *"Wiley Survey of Instrumentation and Measurement"*, John Wiley and Sons, 2001

[115] P. Chow, R. Nobes, T. Kubota, T. Namiki *"Advanced Methods for Electromanetic Simulation"*, Fujitsu Scientific & Technical Journal, 2007

[116] D.A. Mlynski *"Elektrodynamik"*, Lecture Script, Universität Karlsruhe, 1999

[117] Agilent Technologies *"Powering Collaboration and Innovation in the Simulation Design Flow"*, Agilent EEsof Design Forum, 2010

[118] Y. Huang, K. Boyle *"Antennas: From Theory to Practice"*, John Wiley & Sons, 2008

[119] D.B. Davidson *"Computational Electromagnetics for RF and Microwave Engineering"*, Cambridge University Press, 2010

[120] A. Taflove, S. Hagness *"Computational Electrodynamics"*, Artech House, 2005

[121] U. Jakobus *"Comparison of different techniques for the treatment of lossy dielectric/magnetic bodies within the method of moments formulation"*, AEÜ International Journal of electronics and communications, Vol. 54, No. 3, pp. 163-173, 2000

[122] EM Software and Systems SA (Pty) Ltd, *"Feko, quarterly"*, 2005

[123] E.S. Li, K. Sarabandi, *"Low Grazing Incidence Millimeter-Wave Scattering Models and Measurements for Various Road Surfaces"*, IEEE Transactions on Antennas and Propagation, Vol. 47, No. 5, May 1999

[124] K. Holliger, *"Elektromagnetische Verfahren in der Ingenieurgeophysik"*, Institut für Geophysik, ETH-Hoenggerberg, HPP O6, CH-8093 Zürich

[125] N. Geng, W. Wiesbeck, *"Planungsmethoden für die Mobilkommunikation"*, Springer Verlag, 1998

[126] L. Burgess, *"Estimating RKE System Performance"*, Microwaves & RF, 2006

[127] R. Sobot, *"Wireless Communication Electronics: Introduction to RF Circuits and Design Techniques"*, Springer, 2012

[128] M.C. Huynh, W. Stutzman, *"Ground plane effects on planar inverted-F antenna (PIFA) performance"*, IEEE Proceedings, Microwaves, antennas and Propagation, pp.209-213, 2003

[129] D.B. Davidson, *"Computational Electromagnetics for RF and Microwave Engineering"*, Cambridge University Press, 2005

[130] G.-A.-Chakam, *"An Inductively Loaded PIFA for Automotive Applications"*, European Microwave Week 2004, Amsterdam

[131] RoSPA *"Road safety Information"*, The Royal Society for the Prevention of Accidents, RoSPA House, Edgbaston Park, 353 Bristol Road, Birmingham, URL: http://www.rospa.co.uk, 2008

[132] W.C. Vetter *"Run Flat Tires"*, BellaOnline's Cars Editor, BellaOnline, 2008

[133] S. Stankowski, A. Kessi, O. Becheiraz, K. Meier-Engel, M. Meier *"Low frequency magnetic fields in cars, induced by tire magnetisation"*, URL: https://prof.hti.bfh.ch, 2010

[134] F. Iseli, P. Fankhauser, K. Meier-Engel *"Reifenentmagnetisierung Labormessbericht - Kohlenstoffanalyse und Entmagnetisierungsmethoden"*, Fachrichtung Elektrotechnik, Berner Fachhochschule Hochschule für Technik und Informatik HTI Biel, 2006

[135] Continental Automotive GmbH, *"Tyre Basics Passenger Car Tyres"*, Technical Literature, Continental, 2008/2009

[136] Automotive Electronics, URL: http://www.automotive-electronics.co.uk, 2008

A Additional General Information on Tyre Classification

Annex A gathers all additional, accompanying information required for tyre classification process. Firstly possible classification criteria are presented, followed by basic information about tyre annotations and explanation of such therms as speed and load index. Based on this information tabelarized overview for conventional and SSR tyres is presented. Finally representative tyres for further analysis are chosen and listed.

Classification describes the possibility to arrange the subject of matter into certain groups with predefined characteristics. Annex A summarizes all the effort that was needed from statistics to mechanical, geometric and structural point of view in order to make tyre classification possible. Firstly upon analysis of all these themes a correct choice of representative tyres for electromagnetic investigation was possible.

A.1 Standardized Tyre Classification Criteria

Tyres available on the market are very well labelled products, classified upon geometrical construction, size, maximum rotation speed, load index, tread profile and usage (winter, summer or all season tyres). Overall there are twelve major information blocks upon which tyres are described and recognized. Fig. A.1 depicts a typical tyre. Explanation for all information blocks is gathered summarized in Table A.1 and Table A.2.

Figure	Code on the tyre	Explanation
A.1 (1)	Continental	Manufacturer or brand name
A.1 (2)	ContiPremiumContact 2	Product name
A.1 (3)	205 / 55 R 16	Tyre size and construction type
A.1 (4)	Radial Tubeless	Tyre construction type
A.1 (5)	91V	Load and speed index (see Table A.3)
A.1 (12)	SSR	Tyre type: Self Supporting Runflat

Table A.1: Tyre description - physical description

First information to find on a tyre is the manufacturer with his brand name Fig. A.1 (1). Further each tyre has got a specific product name Fig. A.1 (2) which normally is related to tyre usage.

Figure	Code on the tyre	Explanation
A.1 (6)	E4	Compliance with ECE regulations
A.1 (7)	DOT	Compliance with safety standards in USA
A.1 (8)	0227293-s	Approval number of ECE regulation
A.1 (9)	2206	Production date
A.1 (10)	TWI	Tyre Wear Index
A.1 (11)	Made in Germany	Country of production

Table A.2: Tyre description - regulations and standards

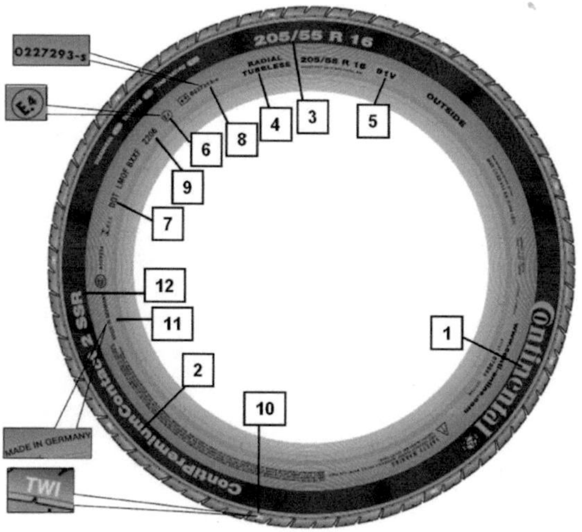

Figure A.1: Tyre labels: (1) Brand, (2) Name, (3) Size, (4) Tube type, (5) Tyre load and speed index, (6) Compliance with ECE regulations, (7) Department of Transportation, (8) Approval number acc. to relevant ECE regulation, (9) Production code, (10) Tread Wear Indicator, (11) Country of origin, (12) Special designation for SSR runflat tyres

Third important parameter is the tyre size Fig. A.1 (3). In presented example it is 205 / 55 R 16. This means that the tyre has width of 205 mm. Side wall height is expressed as percentage of the width, in this case height is equal 55 % of 205 mm which results in 112.75 mm. Height - width ratio between 55 % and 65 % is standard for many vehicles today, whereas modern tyres for high performance cars are having height - width ratio as low as 25 % [135]. Rim diameter is expressed always in inch, in this case 16 inch which corresponds to 406.4 mm (1 inch = 25.4 mm). Letter R corresponds to tyre construction type Fig. A.1 (4). Tyres may have either radial or diagonal construction. R stands for radial construction. Presented tyre is a tubeless Fig. A.1 (4) tyre. Typical vehicle tyres are tubeless nowadays with exception of some special, heavy duty machines and bicycles. Tyre load (Fig. A.1 (5)) in example is 91 which corresponds to 615 kg. Speed index (Fig. A.1 (5)) is to be read from the Table A.3. Letter V corresponds to max 240 km / h. Last marking that reveals physical description is the information that the tyre belongs to construction type namely Self Supporting Runflat (SSR) tyre Fig. A.1 (12).

All further markings correspond to international regulations and not any more to physical description. Inscription indicating compliance with ECE (Economic Commission for Europe) regulations Fig. A.1 (6) indicates the country of homologation (E4 = the Netherlands).

The tyre safety standards for the competent US Department of Transportation is represented by Fig. A.1 (7). Fig. A.1 (8) is an approval number acc. to relevant ECE regulation. Production code Fig. A.1 (9) reveals production date "2206" means 22nd week of year 2006.

Apart from the information how old the tyre is, is the Tread Wear Indicator (TWI) Fig. A.1 (10). Further marking indicates where the tyre was produced Fig. A.1 (11).

A.2 Tyre Speed and Load Index

Tyre speed index defines the maximum speed that a properly inflated tyre may sustain without a risk of uncontrolled puncture or increased wear off. The load index on a passenger car tyre is a numerical code stipulating the maximum load (mass, or weight) each tyre can carry. Index explanation is to be found in Table A.3. Compilation of all tyres available on the market was done upon speed index and tyre size - see Chapter A.3 to Chapter A.6.

The analysis showed very clearly that the present tyres feature applicability for higher speeds and low speed indexes are not popular. Most conventional summer and winter tyres have got speed index T or H which reflects to the maximum speed of 190 km/h to 200 km/h. SSR tyres permit diving at higher speeds over 200 km/h with W, ZR and Y speed index.

Speed Index	Max. Speed	Load Index	kg	Load Index	kg	Load Index	kg	Load Index	kg
M	130 km/h	50	190	69	325	88	560	107	975
P	150 km/h	51	195	70	335	89	580	108	1000
Q	160 km/h	52	200	71	345	90	600	109	1030
R	170 km/h	53	206	72	355	91	615	110	1060
S	180 km/h	54	212	73	365	92	630	111	1090
T	190 km/h	55	218	74	375	93	650	112	1120
U	200 km/h	56	224	75	387	94	670	113	1150
H	210 km/h	57	230	76	400	95	690	114	1180
V	240 km/h	58	236	77	412	96	710	115	1215
W	270 km/h	59	243	78	425	97	730	116	1250
Y	300 km/h	60	250	79	437	98	750	117	1285
ZR	> 240 km/h	61	257	80	450	99	775	118	1320
		2	265	81	462	100	800	119	1360
		63	272	82	475	101	825	120	1400
		64	280	83	487	102	850	121	1450
		65	290	84	500	103	875	122	1500
		66	300	85	515	104	900	123	1550
		67	307	86	530	105	925	124	1600
		68	315	87	545	106	950		

Table A.3: Tyre speed index with corresponding maximum speed and tyre load index with corresponding weight

A.3 Overview of Conventional Summer Tyres

Fig. A.2 comprises all conventional summer tyres available on the European market in 2007. Tyres from selected manufacturers, having biggest market shares, are taken into consideration: Continental, Goodyear, Pirelli, Michelin, Dunlop and Bridgestone.

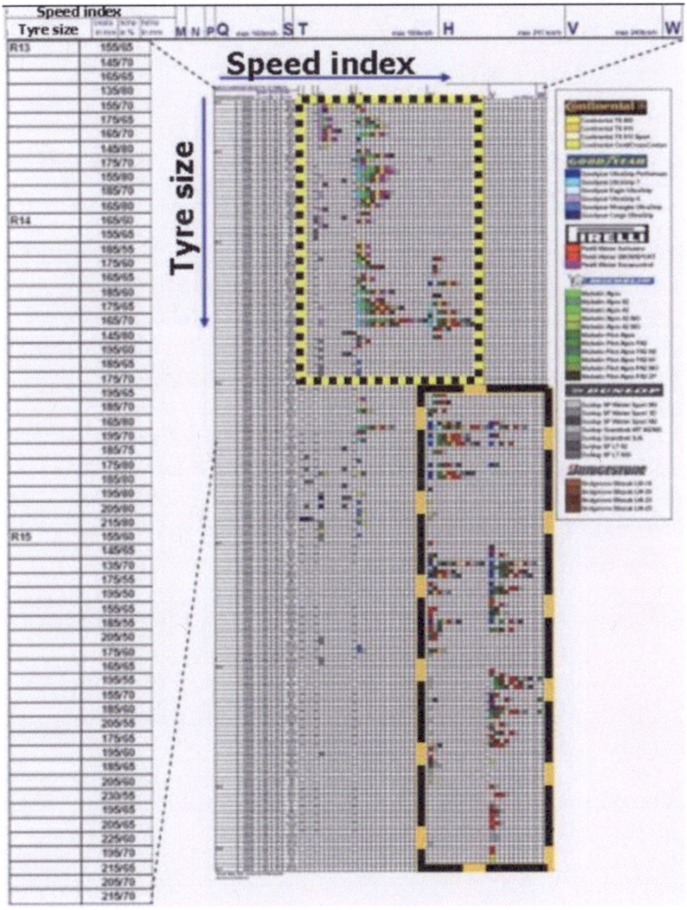

Figure A.2: Compilation of all conventional summer tyres available on the market in 2007 - selected manufacturers

A.4 Overview of Conventional Winter Tyres

Fig. A.3 comprises all conventional winter tyres available on the European market in year 2007. Tyres from selected manufacturers, having biggest market shares, are taken into consideration: Continental, Goodyear, Pirelli, Michelin, Dunlop and Bridgestone.

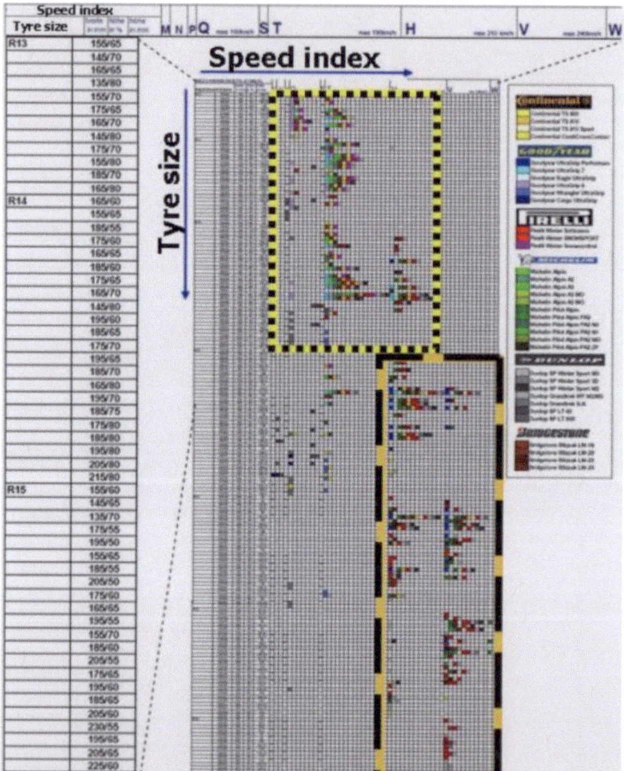

Figure A.3: Compilation of all conventional winter tyres available on the market in 2007 - selected manufacturers

A.5 Overview of SSR Summer Tyres

Fig. A.4 comprises all SSR summer tyres available on the European market in year 2007. Tyres from selected manufacturers, having biggest market shares, are taken into consideration: Continental, Goodyear, Pirelli, Michelin, Dunlop and Bridgestone.

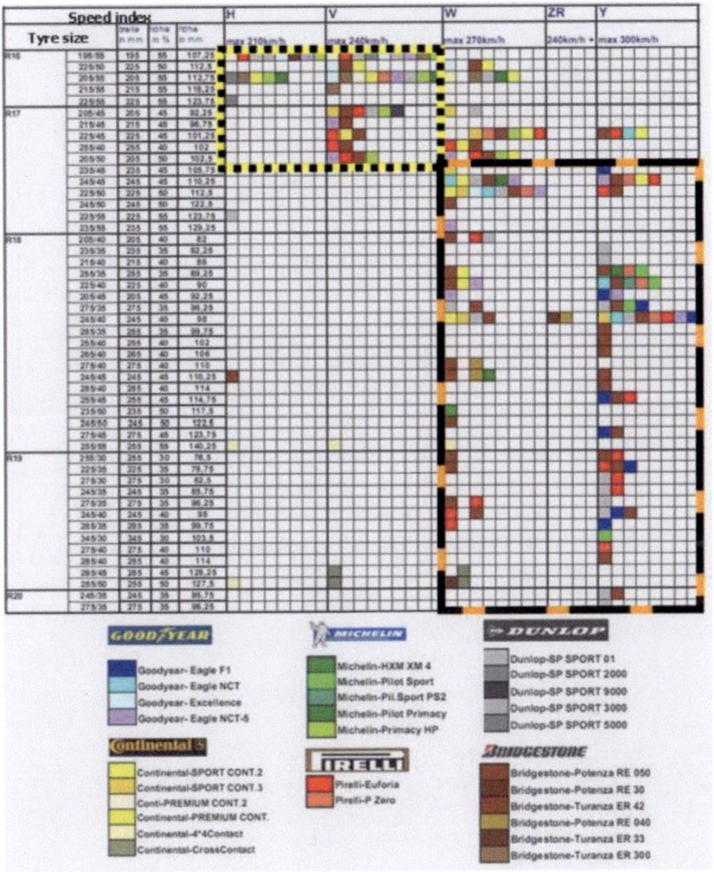

Figure A.4: Compilation of all SSR summer tyres available on the market in 2007 - selected manufacturers

A.6 Overview of SSR Winter Tyres

Fig. A.5 comprises all SSR winter tyres available on the European market in year 2007. Tyres from selected manufacturers, having biggest market shares, are taken into consideration: Continental, Goodyear, Pirelli, Michelin, Dunlop and Bridgestone.

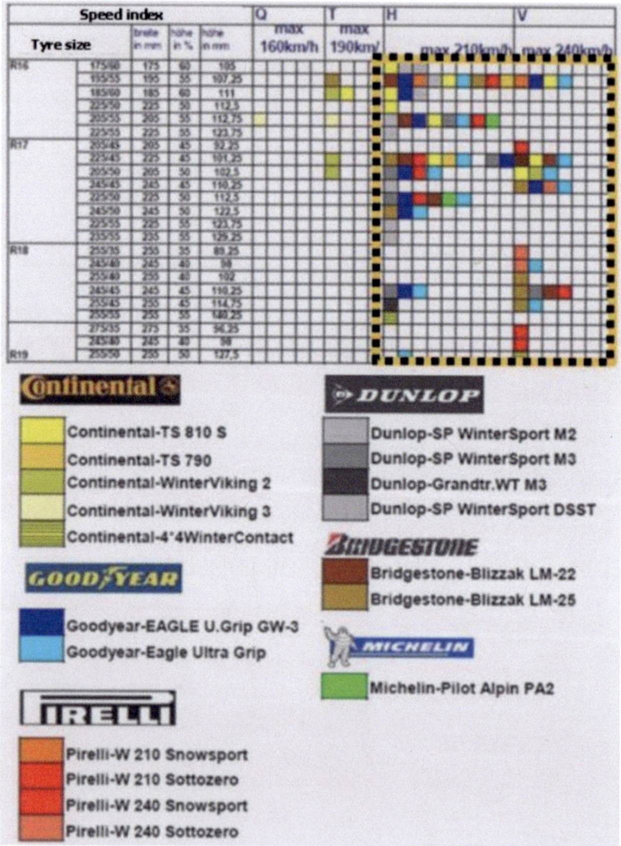

Figure A.5: Compilation of all SSR winter tyres available on the market in 2007 - selected manufacturers

A.7 List of Conventional and SSR Tyres Chosen for Analysis

Tyres chosen for analysis are listed in Table A.4.

	Summer Tyres		Winter Tyres	
	Conventional 195/65 R15		Conventional 195/65 R15	
Manufacturer	Product Name	Speed Index	Product Name	Speed Index
Continental	EcoContact 3	T	TS 810	T
GoodYear	Vector 5+	T	UltraGrip 7	T
Pirelli	P6	T	Snow Sport 190	T
Michelin	Energy E3A	T	Alpin A3	T
Dunlop	SP 30	T	Winter Sport 3D	T
Bridgestone	B250	T	Blizzak LM-25	T
	Conventional 245/40 R18		Conventional 245/40 R18	
Manufacturer	Product Name	Speed Index	Product Name	Speed Index
Continental	Sport Contact 2	V	Winter Contact TS810	V
GoodYear	Eagle F1 Assymetric	Y	Ultra Grip	V
Pirelli	P Zero Rosso	Y	Sottozero	V
Michelin	Piolot Sport	Y	Pilot Alpin	V
Dunlop	Sport 01	Y	Winter Sport M3	V
Bridgestone	Potenza RE050	Y	*	
	SSR 245/40 R18		SSR 245/40 R18	
Manufacturer	Product Name	Speed Index	Product Name	Speed Index
Continental	Sport Contact 3	W	*	
GoodYear	Eagle NCT	Y	Eagle Ultra Grip	V
Pirelli	Euforia	Y	Sottozero W240	V
Michelin	*		*	
Dunlop	SP Sport 01	Y	SP Winter Sport M3	V
Bridgestone	Potenza RE 040	ZR	*	

Table A.4: Overview of conventional and SSR tyres chosen for further analysis (*Tyres not available on market at the time of the investigation)

B Radiation Pattern Measurement

Annex B complements the thesis with information on radiation pattern measurement setup as well as detailed numerous measurement results accompanying tyre characterization process. Chapter 3 and Chapter 5 refer hereto.

Complementary to Chapter 3 characteristic measured radiation pattern of the chosen conventional and SSR tyres as well as average Gain values are presented. Further, complementary to Chapter 5 and Chapter 6 detailed results are presented of the characteristic measured and simulated values for:

- Gain of the stand alone Wheel Units.
- Gain of the Wheel Units on the rim.
- Gain of the Wheel Units on the rim with representative tyres.

The antenna radiation pattern is the information on the radiation properties of the antenna as a function of a spherical coordinate system (Fig. B.1).

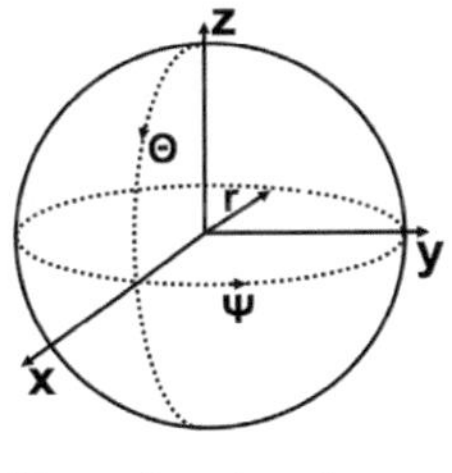

(a) Definition of rectangular coordinate system (x,y,z) and spherical coordinate system (r, Θ, Ψ).

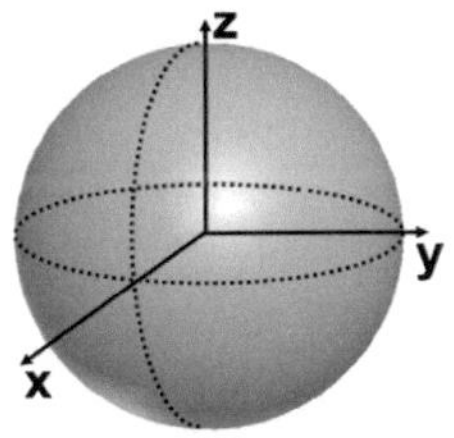

(b) Isotropic radiation characteristic

Figure B.1: Definition of coordinate systems and isotropic radiation pattern

Radiation pattern describes spatial radiation behaviour. An isotropic radiator, which exhibits the same radiation in any direction form the electromagnetic source, depicted in Fig. B.1b, is not possible to be realized and is set as a comparison standard [39].

For radiation pattern characterization beneath, presented in Fig. B.2, orientation in the coordinate system and principal planes definition was set. This is the reference for all simulations and measurements of radiation pattern characteristics in the whole document.

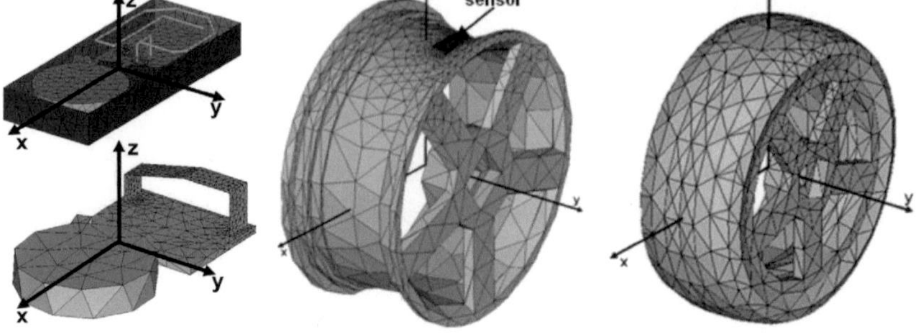

(a) Wheel Unit with IFA / loop (b) Wheel Unit on the rim (c) Wheel Unit on the rim with tyre

Figure B.2: Measurement and simulation coordinate system

B.1 Radiation Pattern Measurement Setup

The here described radiation pattern measurement setup for any Device / Material Under Test (DUT / MUT) and the orientation (positioning) of the DUT is valid for all achieved results in the thesis. Radiation pattern characterization was conducted in an anechoic chamber. The measurement setup consisted of receiver antenna, test receiver and a computer controlled turn table with DUT. Turn table rotation was controlled over 360° with resolution of 2°.

MUT may be any object, that radiates electromagnetic energy. In case of this thesis these are such antenna modules as WU and receiver, bigger objects as WU on the rim or in tyre cavity. Measurement setup with tyre as MUT is visible in Fig. B.3.

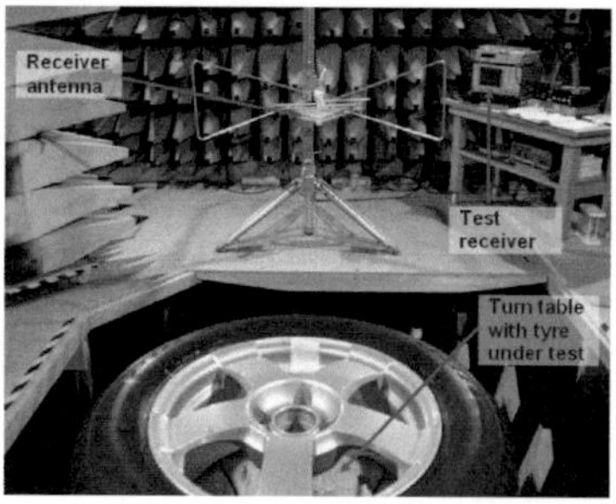

Figure B.3: Radiation pattern measurement setup

The corresponding measurement block diagramm of the described measurement setup is given in Fig. B.4.

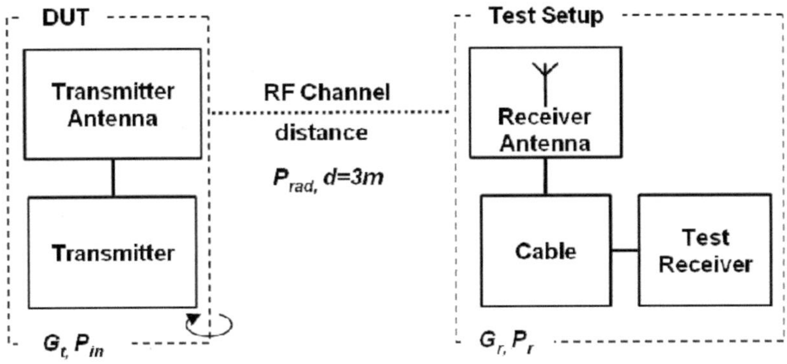

Figure B.4: Block diagram of Two-Antenna-Method measurement setup

Both parts, system under investigation (DUT) and the measurement chamber infrastructure (Test Setup), are linked by RF-Channel. In Two-Antenna-Method [14] all relevant measurement and system parameters are well defined and used for the post-processing of the acquired characterization data.

B.2 Radiation Pattern Influenced by Conventional Tyres - Detailed Results

Annex B.2 gathers complementary results to Chapter 3 where radiation pattern influenced by different conventional tyre brands was discussed.

Measured radiation patterns are depicted respectively:

- for small (195/65 R15) summer tyres in Fig. B.5
- for small (195/65 R15) winter tyres in Fig. B.6
- for big (245/40 R18) summer tyres in Fig. B.7
- for big (245/40 R18) winter tyres in Fig. B.8.

Already after the first evaluation comparison it may be stated that the radiation pattern shape is the same for every measured tyre of the same size. What is readable from amplitude differences, is that tyre of every manufacturer reveals its specific damping. It is however important to note, that taking into account typical measurement accuracy in anechoic chamber of 3 dB, these differences are rather small.

The best (with the highest radiation amplitude) small, summer tyre is the Continental Eco Contact, whereas the worst (with the lowest radiation amplitude) tyre seems to be Bridgestone B250. Maximum measured gain difference between the average values is 4.7 dB (Table B.1). The best small, winter tyre is the Bridgestone Blizzak LM 25. The worst small, winter tyre is Michelin Alpin A3. Maximum difference in radiation pattern between these two tyres amounts 5 dB (Table B.2).

Analysing the results from tyres that were classified as big tyres, the best summer tyre is the Pirelli P Zero Rosso and best winter tyre is the Pirelli Sottozero. The worst summer tyre is the Michelin Piolot Sport and winter tyre is the Continental Winter Contact. Maximum gain difference between summer tyres amounts 3.9 dB (Table B.3), and between winter tyres 2 dB (Table B.4).

B.3 Radiation Pattern Values for SSR Tyres - Detailed Results

Annex B.3 gathers complementary results to Chapter 3 where radiation pattern influenced by different SSR tyre brands was discussed.

Table B.6 features characteristic measured radiation pattern values for SSR, (245/40 R18) summer tyres and Table B.5 features characteristic measured radiation pattern values for SSR, (245/40 R18) winter tyres.

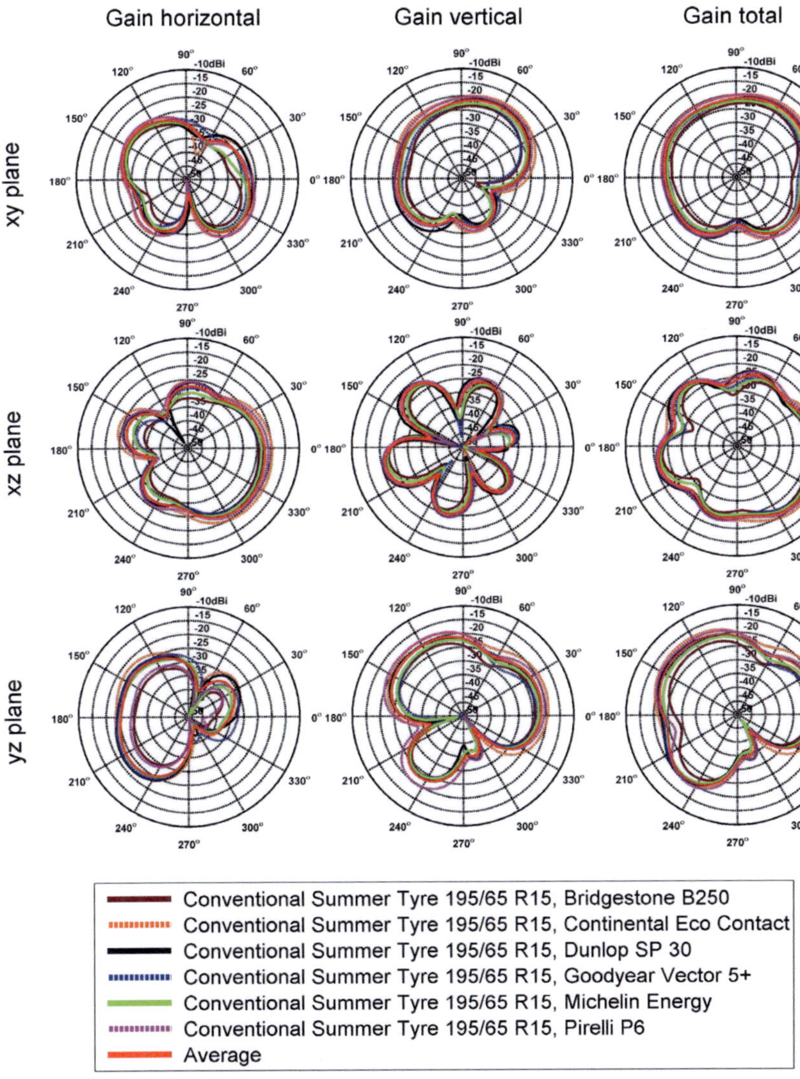

Figure B.5: Measured radiation pattern of conventional small (195/65 R15) summer tyres

meas. planes	Tyre name	Gain horizontal in dBi			Gain vertical in dBi			Gain total in dBi		
	summer tyres 195/65 R15	max	min	avg.	max	min	avg.	max	min	avg.
xy plane	Bridgestone B250	-26.4	-51.2	-30.6	-22.7	-44.5	-27.9	-22.4	-34.8	-26
	Continental Eco Contact	-23.8	-50	-26.8	-18.7	-38.1	-23.6	-18.5	-31.2	-21.9
	Dunlop SP 30	-24.9	-47.9	-27.9	-19.9	-37.2	-24.8	-19.6	-34.4	-23.1
	Goodyear Vector 5+	-25	-44.4	-28.6	-21.9	-40.9	-27	-21.3	-34.3	-24.7
	Michelin Energy	-25.4	-48.5	-29.2	-21.6	-41.9	-26.3	-21.4	-32.8	-24.5
	Pirelli P6	-23.4	-50.5	-26.9	-19.3	-38.4	-24.2	-19.2	-31.7	-22.4
	average	-25	-44.8	-28.2	-20.4	-38.1	-25.4	-20.3	-32.8	-23.6
xz plane	Bridgestone B250	-23	-38.2	-27.7	-22.9	-49.4	-28.5	-22.1	-33.1	-25.1
	Continental Eco Contact	-18.9	-27.5	-23.2	-20.9	-52.3	-26.5	-18.5	-26.7	-21.6
	Dunlop SP 30	-21.4	-47.7	-25.6	-20.3	-48.3	-26	-20.1	-28.3	-22.8
	Goodyear Vector 5+	-22.6	-37.6	-26.8	-21.3	-50.2	-26.8	-18.2	-30.5	-23.8
	Michelin Energy	-22.2	-28.1	-26.7	-22.1	-54	-27.7	-21.3	-29.8	-24.2
	Pirelli P6	-20.2	-39.2	-24.3	-20.5	-47.9	-26.1	-19.5	-26.4	-22.1
	average	-21.3	-44.9	-26.9	-21.1	-37.3	-25.4	-20.4	-27.7	-23.1
yz plane	Bridgestone B250	-22.8	-49.6	-26.1	-28.3	-54.9	-33	-22.1	-46.8	-25.3
	Continental Eco Contact	-18.5	-49.3	-23	-21.7	-48.6	-26.1	-17.2	-45.2	-20.6
	Dunlop SP 30	-20.4	-49.2	-24.1	-21.8	-47.6	-26.9	-18.7	-43.5	-22.3
	Goodyear Vector 5+	-21.9	-52.3	-25.6	-22.1	-42.6	-27	-20.1	-39.9	-23.3
	Michelin Energy	-21.8	-55.4	-25.1	-23.1	-55.7	-28.3	-20.5	-47.5	-23.4
	Pirelli P6	-18.6	-51	-22	-27.2	-52.8	-31.9	-18.1	-43.3	-21.5
	average	-20.4	-44.3	-23.8	-23.3	-46.2	-28.2	-19.3	-42.9	-22.5

Table B.1: Characteristic measured radiation pattern values for conventional, small (195/65 R15) summer tyres

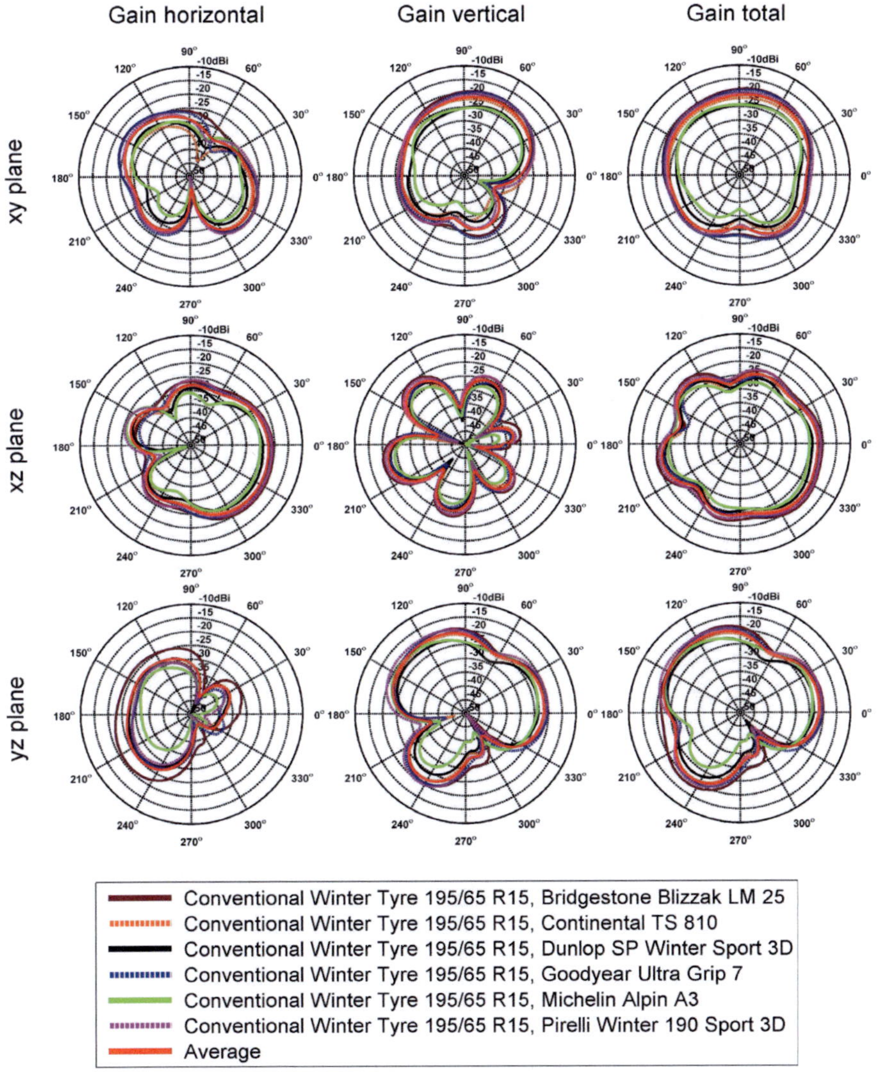

Figure B.6: Measured radiation pattern of conventional small (195/65 R15) winter tyres

meas. planes	Tyre type	Gain horizontal in dBi			Gain vertical in dBi			Gain total in dBi		
	winter tyres 195/65 R15	max	min	avg.	max	min	avg.	max	min	avg.
xy plane	Bridgestone Blizzak LM 25	-22.6	-41.8	-26.3	-18.8	-40.6	-23.8	-18.2	-28.4	-21.8
	Continental TS 810	-25.6	-53.3	-29.2	-21.5	-37.4	-25.7	-21.5	-34	-24.1
	Dunlop SP Winter Sport 3D	-26.9	-56.5	-30.5	-23.5	-43.5	-28.3	-23.3	-34.1	-26.3
	Goodyear Ultra Grip 7	-22.8	-46.7	-26.4	-18.8	-42.8	-24	-18.6	-28	-22
	Michelin Alpin A3	-27.9	-54.8	-31.1	-23.9	-44.2	-28.8	-23.5	-37.6	-26.8
	Pirelli Winter 190 Sport 3D	-23.4	-50.5	-26.9	-19.3	-38.4	-24.2	-19.2	-31.7	-22.4
	average	-24.8	-45.7	-28	-20.5	-38.3	-25.4	-20.1	-32.7	-23.5
xz plane	Bridgestone Blizzak LM 25	-19.6	-42.2	-25.4	-19.5	-33.3	-24	-18.8	-29.1	-21.6
	Continental TS 810	-21.6	-49.9	-27.4	-21.3	-37	-25.7	-20.7	-28.7	-23.5
	Dunlop SP Winter Sport 3D	-22.4	-50.1	-28.1	-23	-47.2	-27.2	-21.9	-29.3	-24.6
	Goodyear Ultra Grip 7	-22.2	-49.3	-27.8	-20.7	-34.7	-25	-20	-30.5	-23.2
	Michelin Alpin A3	-23.8	-56.4	-29.6	-23.8	-48.4	-27.6	-23.1	-30.7	-25.5
	Pirelli Winter 190 Sport 3D	-20.5	-47.9	-26.1	-20.2	-39.2	-24.3	-19.5	-26.4	-22.1
	average	-21.5	-46	-27.2	-21.2	-36.6	-25.4	-20.5	-27.8	-23
yz plane	Bridgestone Blizzak LM 25	-24.7	-45.3	-25.9	-18.8	-50	-22.6	-17.9	-41.9	-21
	Continental TS 810	-25.8	-56.5	-30.7	-20.8	-54.3	-23.9	-20.2	-45.5	-23.1
	Dunlop SP Winter Sport 3D	-26.2	-50.3	-30.9	-22.4	-53.9	-25.7	-21.3	-46.8	-24.5
	Goodyear Ultra Grip 7	-26.2	-50.7	-31.1	-18.9	-43.6	-22.5	-18.4	-40.5	-22
	Michelin Alpin A3	-29.6	-67.4	-34.9	-20.4	-42.6	-25	-19.9	-41.5	-24.6
	Pirelli Winter 190 Sport 3D	-27.2	-52.8	-31.9	-18.6	-51	-22	-18.2	-43.2	-21.5
	average	-20	-39.8	-23.4	-25.1	-45.7	-30	-19.4	-39.2	-22.6

Table B.2: Characteristic measured radiation pattern values for conventional, small (195/65 R15) winter tyres

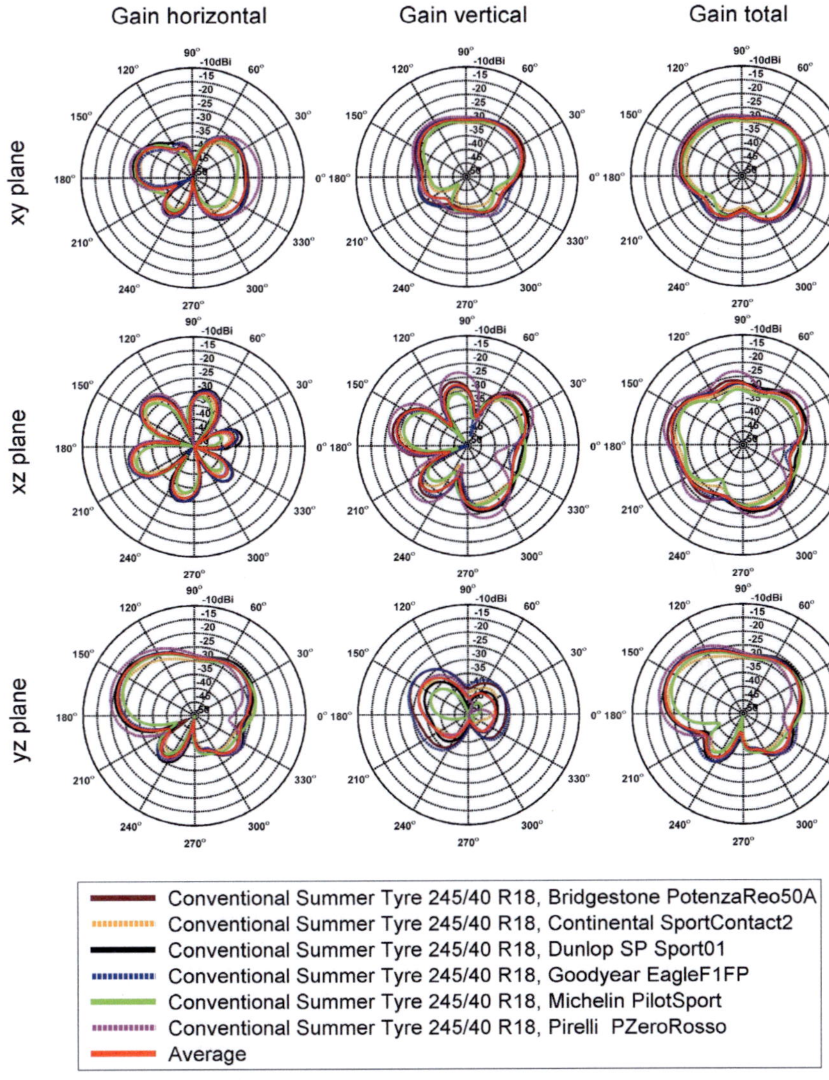

Figure B.7: Measured radiation pattern of conventional big (245/40 R18) summer tyres

meas. planes	Tyre type	Gain horizontal in dBi			Gain vertical in dBi			Gain total in dBi		
	summer tyres 245/40 R18	max	min	avg.	max	min	avg.	max	min	avg.
xy plane	Bridgestone Potenza Reo 50A	-29.1	-56.4	-32.7	-26.4	-38.7	-30.3	-25.1	-38	-28.3
	Continental Sport Contact2	-30.6	-55.8	-35	-27.1	-46.2	-31.9	-26	-38.7	-30.2
	Dunlop SP Sport01	-28.1	-54.4	-32.4	-26.9	-37.7	-30.7	-25.4	-37.4	-28.5
	Goodyear Eagle F1FP	-29.6	-59.4	-32.9	-27.1	-37.9	-30.5	-25.5	-37.2	-28.5
	Michelin Pilot Sport	-28.9	-54.5	-34.2	-27.7	-46.2	-32.1	-26.3	-36.9	-30
	Pirelli P Zero Rosso	-26.1	-49.2	-30.6	-26.3	-36.6	-30.3	-24.6	-36.3	-27.4
	average	-28.9	-55	-33	-26.9	-41.1	-31	-25.5	-37.4	-28.9
xz plane	Bridgestone Potenza Reo 50A	-26	-57	-31.7	-22.3	-47.6	-27.6	-22.3	-32.6	-26.2
	Continental Sport Contact2	-27.8	-56.3	-33.7	-24.2	-43	-29.7	-24	-32.5	-28.3
	Dunlop SP Sport01	-25.7	-53.2	-31.2	-23.4	-50.7	-28	-23.1	-30.3	-26.3
	Goodyear Eagle F1FP	-25.9	-57.1	-31.5	-23.8	-54.6	-28.8	-23.6	-34.1	-27
	Michelin Pilot Sport	-27.9	-62	-34.2	-25.9	-48.1	-30.1	-25	-35	-28.7
	Pirelli P Zero Rosso	-25.6	-58	-31.9	-20.5	-43.2	-25.7	-20.5	-39	-24.8
	average	-26.6	-58.1	-32.6	-23.3	-47.3	-28.4	-23.1	-34.6	-27
yz plane	Bridgestone Potenza Reo 50A	-20.5	-52.2	-27	-27	-46.6	-33.4	-19	-42.3	-26.1
	Continental Sport Contact2	-22.4	-55.5	-28.3	-31	-48.3	-37.6	-21.8	-45.8	-27.8
	Dunlop SP Sport01	-20	-43	-26.3	-32.5	-47.7	-38	-19.7	-40.6	-26
	Goodyear Eagle F1FP	-20.9	-51.3	-27.2	-26.8	-48.4	-32.8	-19.9	-42	-26.1
	Michelin Pilot Sport	-22	-59	-28	-35.1	-52.1	-42.6	-21.8	-48.9	-27.8
	Pirelli P Zero Rosso	-18.5	-48.5	-25.5	-30.1	-50.6	-37.2	-18.2	-43.2	-25.2
	average	-20.9	-53.3	-27.2	-30	-49.2	-36.7	-20.1	-44.4	-26.6

Table B.3: Characteristic measured radiation pattern values for conventional, big (245/40 R18) summer tyres

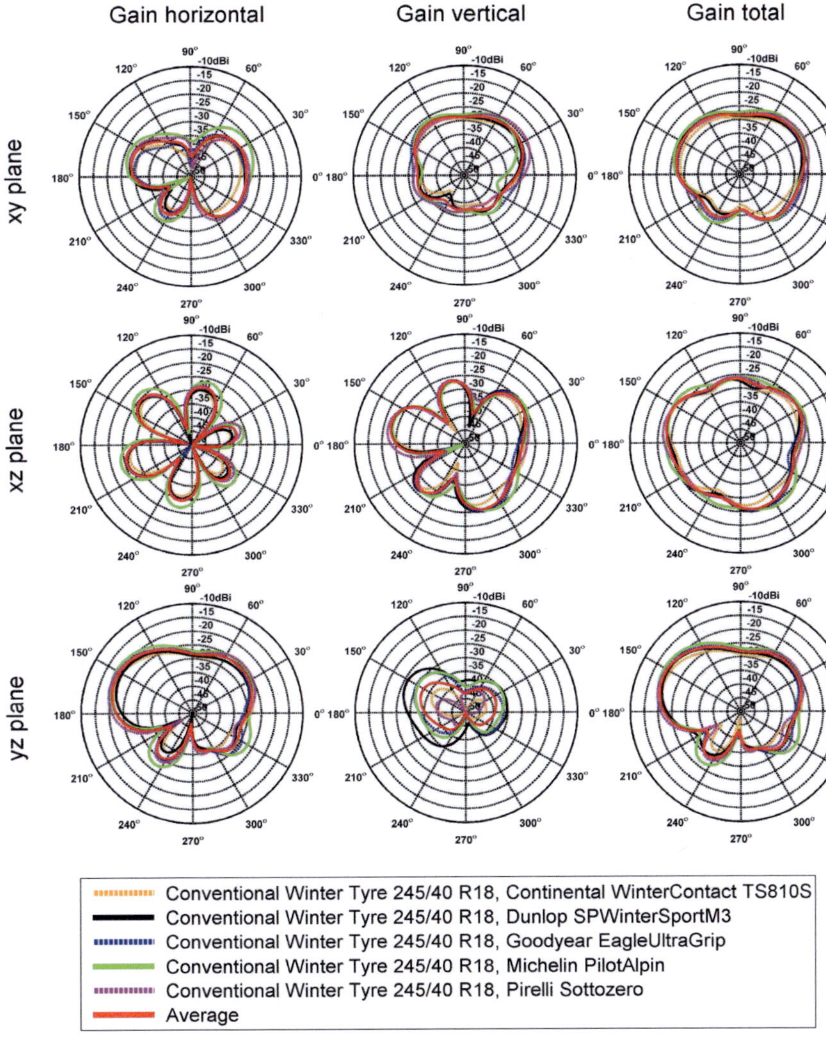

Figure B.8: Measured radiation pattern of conventional big (245/40 R18) winter tyres

meas. planes	Tyre type	Gain horizontal in dBi			Gain vertical in dBi			Gain total in dBi		
	winter tyres 245/40 R18	max	min	avg.	max	min	avg.	max	min	avg.
xy plane	Continental Winter Contact	-29.6	-53.4	-33.9	-26.7	-42.4	-31.2	-25.3	-38.6	-29.3
	Dunlop SP Winter Sport M3	-27.5	-56.7	-32	-26.9	-41.6	-30.7	-25.3	-36.6	-28.3
	Goodyear Eagle Ultra Grip	-28.5	-52.7	-32.3	-26.1	-37.9	-29.6	-24.8	-36.9	-27.7
	Michelin Pilot Alpin	-28	-52.6	-31.8	-25.5	-35.3	-29.2	-24	-35.1	-27.3
	Pirelli Sottozero	-27.5	-54.1	-31.5	-24.1	-38	-29.4	-23	-37	-27.3
	average	-28.2	-53.9	-32.3	-25.9	-39.1	-30	-24.5	-36.8	-28
xz plane	Continental Winter Contact	-26.5	-54.6	-32.5	-22.8	-44.8	-28.5	-22.7	-31	-27.1
	Dunlop SP Winter Sport M3	-25.9	-53.4	-31.7	-22.4	-44	-27.5	-22.5	-30.8	-26.2
	Goodyear Eagle Ultra Grip	-25.4	-62.5	-31.1	-22.6	-49.2	-27.5	-22.4	-32	-25.9
	Michelin Pilot Alpin	-24.7	-53.3	-30.6	-24	-50.8	-28	-23.2	-29.2	-26.1
	Pirelli Sottozero	-25.5	-49	-31	-21.6	-42.8	-27.2	-21.6	-29.1	-25.7
	average	-25.6	-54.6	-31.4	-22.7	-46.3	-27.8	-22.5	-30.4	-26.2
yz plane	Continental Winter Contact	-20.4	-52.2	-26.8	-36.5	-59.1	-43.2	-20.3	-47.7	-26.7
	Dunlop SP Winter Sport M3	-20.8	-50.3	-27.2	-26.1	-46.1	-32.8	-19.7	-41.1	-26.2
	Goodyear Eagle Ultra Grip	-19.3	-43.3	-25.8	-29.6	-51.7	-36	-18.9	-42.2	-25.5
	Michelin Pilot Alpin	-18.9	-45.3	-25.3	-46.6	-62.4	-50	-18.9	-42.4	-25.3
	Pirelli Sottozero	-18.7	-45.8	-25.2	-39.2	-64.4	-43.8	-18.6	-42.3	-25.1
	average	-19.6	-47.4	-26.1	-35.6	-56.7	-41.1	-19.3	-43.1	-25.8

Table B.4: Characteristic measured radiation pattern values for conventional, big (245/40 R18) winter tyres

meas. planes	Tyre type	Gain horizontal in dBi			Gain vertical in dBi			Gain total in dBi		
	RoF winter tyres 245/40 R18	max	min	avg.	max	min	avg.	max	min	avg.
xy plane	Dunlop Wintersoprt	-28.1	-56.6	-33.3	-24.7	-50.3	-30.4	-23.9	-35.8	-28.6
	Goodyear Eagle Ultra Grip	-28.2	-50.6	-32.5	-26	-50.7	-31.2	-24.3	-36.9	-28.8
	Pirelli Sottozero	-28.2	-52.1	-32.8	-23.6	-48.2	-28.6	-22.6	-34	-27.2
	average	-28.2	-51.3	-32.9	-24.7	-46.2	-30.0	-23.5	-35.4	-28.2
xz plane	Dunlop Wintersoprt	-22.2	-45.8	-27.8	-25	-57.5	-31.7	-21.9	-29.8	-26.4
	Goodyear Eagle Ultra Grip	-23.7	-43.8	-29.2	-26.2	-55.8	-32.6	-23.6	-30.5	-27.6
	Pirelli Sottozero	-21.8	-49.1	-27.5	-23.2	-60.5	-29.6	-21.5	-31.1	-25.5
	average	-22.6	-42.4	-28.1	-24.8	-49.5	-31.1	-22.3	-30.1	-26.4
yz plane	Dunlop Wintersoprt	-25.5	-63.1	-32.9	-21.7	-53.9	-27.2	-20.2	-41.9	-26.2
	Goodyear Eagle Ultra Grip	-23.4	-45.9	-30.2	-27	-54.9	-30.4	-21.9	-43.3	-27.3
	Pirelli Sottozero	-25.5	-56.1	-32.9	-20	-52.6	-25.4	-18.9	-49.4	-24.7
	average	-24.7	-49.2	-31.8	-22.1	-51.8	-27.3	-20.2	-43.5	-26.0

Table B.5: Characteristic measured radiation pattern values for SSR, (245/40 R18) winter tyres

meas. planes	Tyre type	Gain horizontal in dBi			Gain vertical in dBi			Gain total in dBi		
	RoF summer tyres 245/40 R18	max	min	avg.	max	min	avg.	max	min	avg.
xy plane	Bridgestone Potenza	-32.9	-51.3	-37.9	-29.3	-49.9	-34.1	-27.7	-41.7	-32.6
	Continental Conti Sport Contact	-35.8	-56.6	-41	-28.5	-48.1	-33.2	-27.8	-41.9	-32.6
	Dunlop SP Sport01	-29.6	-49.7	-34.2	-25.4	-50.1	-30.9	-24.3	-37.7	-29.2
	Goodyear NCT5	-29	-52.5	-34.3	-26.7	-59.2	-31.9	-25.6	-36.8	-29.9
	Pirelli Euforia	-32.1	-55.3	-37.2	-26.6	-53.2	-32.1	-26.1	-38.9	-30.9
	average	-31.9	-52.3	-36.2	-27.1	-48.2	-32.3	-26.1	-38.2	-30.8
xz plane	Bridgestone Potenza	-28.5	-38.3	-31.7	-29	-59.7	-34.9	-27.1	-35	-30
	Continental Conti Sport Contact	-29	-42	-33.1	-28.8	-57.3	-34.8	-27.7	-34.9	-30.9
	Dunlop SP Sport01	-23.2	-45.8	-29.2	-26.1	-59.4	-32.5	-23.2	-32.1	-27.6
	Goodyear NCT5	-23.5	-45.6	-28.8	-26.4	-61.4	-32.7	-23.5	-30.7	-27.4
	Pirelli Euforia	-24.7	-43.8	-30.5	-26.5	-63.6	-32.9	-24.6	-33.2	-28.6
	average	-25.2	-38.9	-30.4	-27.4	-57.7	-33.5	-25.0	-31.6	-28.7
yz plane	Bridgestone Potenza	-30	-49.5	-36.7	-28.9	-56.7	-32.2	-26.7	-48.3	-30.8
	Continental Conti Sport Contact	-34.8	-59.1	-42	-27.1	-53.4	-31.2	-26.5	-49.8	-30.9
	Dunlop SP Sport01	-25.8	-59.6	-33.1	-23.9	-54	-29	-21.7	-46.7	-27.6
	Goodyear NCT5	-26	-67.8	-33.4	-23.5	-56.5	-28.7	-21.6	-44.6	-27.4
	Pirelli Euforia	-26.6	-59.4	-33.8	-25.5	-55.7	-30.2	-23	-48.2	-28.6
	average	-27.6	-51.9	-34.9	-25.4	-54.3	-30.0	-23.4	-47.1	-28.8

Table B.6: Characteristic measured radiation pattern values for SSR, (245/40 R18) summer tyres

B.4 Characteristic Measured Radiation Pattern Values - TPMS Components

Annex B.4 gathers characteristic measured radiation pattern values in terms of antenna DUT gain. For simulation and measurement results minimum, maximum and average values were gathered in beneath tables.

B.4.1 Wheel Unit

Measured and simulated radiation pattern values for WUs (with IFA and loop antenna) are gathered in Table B.7 and Table B.8. Both WUs are described in detail: IFA in Chapter 5.3.4 and loop in Chapter 5.3.6.

		Gain horizontal in dBi			Gain vertical in dBi			Gain total in dBi		
meas. plane	Result	max	min	avg.	max	min	avg.	max	min	avg.
xy	Sim.	-16.0	-22.6	-18.4	-24.5	-25.0	-24.8	-15.5	-20.6	-17.5
	Meas.	-15.6	-20.8	-17.4	-22.7	-26.5	-24.2	-14.9	-19.1	-16.6
yz	Sim.	-22.5	-36.4	-25.4	-16.5	-42.2	-19.6	-16.3	-23.2	-18.6
	Meas.	-20.4	-29.3	-22.9	-16.2	-47.3	-19.4	-15.9	-20.5	-17.8
xz	Sim.	-16.0	-17.4	-16.7	-24.5	-51.5	-27.6	-15.4	-17.3	-16.4
	Meas.	-15.3	-17.4	-16.1	-22.8	-40.4	-25.9	-14.9	-17.0	-15.7

Table B.7: Characteristic measured radiation pattern values for stand alone Wheel Unit with Inverted F Antenna

		Gain horizontal in dBi			Gain vertical in dBi			Gain total in dBi		
meas. plane	Result	max	min	avg.	max	min	avg.	max	min	avg.
xy	Sim.	-28.4	-31.8	-29.8	-24.6	-37.7	-27.4	-23.1	-30.7	-25.4
	Meas.	-30.7	-34.8	-32.1	-22.9	-49.0	-25.9	-21.1	-31.1	-23.9
yz	Sim.	-28.4	-31.1	-29.5	-24.5	-24.8	-24.6	-23.1	-23.8	-23.4
	Meas.	-30.4	-42.6	-34.1	-21.9	-22.2	-22.0	-21.4	-22.0	-21.7
xz	Sim.	-23.9	-43.5	-26.9	-29.7	-69.5	-32.9	-23.4	-32.7	-25.9
	Meas.	-22.5	-43.3	-25.4	-35.9	-65.9	-39.0	-21.4	-36.5	-24.2

Table B.8: Characteristic measured radiation pattern values for stand alone Wheel Unit with loop

B.4.2 Wheel Unit on the Rim

Wheel Unit was measured on small (15") and big (18") rim. This Appendix gathers complementary results to Chapter 6.8.2. Measured and simulated radiation pattern values for WUs with IFA and loop antenna with 15" rim are gathered in Table B.9 and Table B.10. Measured and simulated radiation pattern values for WUs with IFA and loop antenna with 18" rim are gathered in Table B.11 and Table B.12.

		Gain horizontal in dBi			Gain vertical in dBi			Gain total in dBi		
meas. plane	Result	max	min	avg.	max	min	avg.	max	min	avg.
xy	Sim.	-29.5	-40.4	-32.9	-23.0	-28.6	-25.5	-22.9	-27.1	-24.8
	Meas.	-30.6	-54.9	-33.9	-24.8	-29.9	-26.2	-24.5	-27.7	-25.5
yz	Sim.	-28.9	-43.2	-32.7	-20.7	-29.2	-24.4	-20.2	-28.9	-23.8
	Meas.	-27.7	-41.7	-30.3	-23.2	-31.3	-26.6	-22.1	-30.3	-25.0
xz	Sim.	-24.5	-39.1	-27.9	-22.3	-45.7	-25.5	-21.2	-26.5	-23.5
	Meas.	-25.7	-40.4	-29.2	-24.0	-46.5	-26.7	-23.1	-27.9	-24.8

Table B.9: Characteristic measured radiation pattern values for Wheel Unit with Inverted F Antenna on the 15" rim (small rim)

		Gain horizontal in dBi			Gain vertical in dBi			Gain total in dBi		
meas. plane	Result	max	min	avg.	max	min	avg.	max	min	avg.
xy	Sim.	-23.1	-56.2	-25.5	-12.9	-53.7	-17.8	-12.9	-24.6	-17.1
	Meas.	-23.9	-40.9	-26.1	-14.4	-56.6	-19.5	-14.3	-25.2	-18.7
yz	Sim.	-37.1	-54.8	-41.6	-12.2	-19.4	-14.7	-12.2	-19.3	-14.7
	Meas.	-34.2	-60.5	-37.7	-12.9	-21.3	-15.9	-12.9	-21.2	-15.8
xz	Sim.	-12.2	-24.9	-16.3	-29.1	-57.0	-34.6	-12.2	-24.5	-16.3
	Meas.	-12.7	-27.3	-17.4	-29.8	-48.0	-34.1	-12.7	-26.3	-17.3

Table B.10: Characteristic measured radiation pattern values for Wheel Unit with loop antenna on the 15" rim (small rim)

		Gain horizontal in dBi			Gain vertical in dBi			Gain total in dBi		
Meas. plane	Result	max	min	avg.	max	min	avg.	max	min	avg.
xy	Sim.	-26.8	-47.1	-31.4	-22.0	-28.3	-24.2	-21.4	-27.3	-23.4
	Meas.	-25.9	-41.7	-30.4	-22.2	-30.1	-24.1	-21.2	-28.2	-23.2
yz	Sim.	-28.9	-43.1	-33.5	-19.8	-32.0	-23.4	-19.6	-28.7	-23.0
	Meas.	-31.3	-41.0	-34.5	-20.3	-30.3	-23.0	-20.2	-28.9	-22.7
xz	Sim.	-23.7	-39.5	-26.7	-21.7	-46.3	-24.9	-20.5	-26.2	-22.7
	Meas.	-22.9	-39.9	-25.9	-21.1	-49.9	-24.6	-20.5	-25.8	-22.2

Table B.11: Characteristic measured radiation pattern values for Wheel Unit with Inverted F antenna on the 18" rim (big rim)

		Gain horizontal in dBi			Gain vertical in dBi			Gain total in dBi		
Meas. plane	Result	max	min	avg.	max	min	avg.	max	min	avg.
xy	Sim.	-23.9	-48.1	-26.5	-15.1	-54.4	-19.6	-15.1	-26.1	-18.8
	Meas.	-22.6	-44.7	-26.6	-17.2	-40.8	-21.9	-17.2	-26.0	-20.6
yz	Sim.	-38.6	-67.7	-43.2	-12.0	-21.7	-15.1	-12.0	-21.6	-15.0
	Meas.	-30.1	-46.5	-33.9	-13.2	-23.7	-16.7	-13.1	-23.4	-16.6
xz	Sim.	-12.1	-30.2	-16.3	-29.9	-54.4	-35.7	-12.1	-29.2	-16.3
	Meas.	-13.1	-34.2	-17.7	-30.3	-54.1	-36.0	-13.1	-29.9	-17.6

Table B.12: Characteristic measured radiation pattern values for Wheel Unit with loop antenna on the 18" rim (big rim)

B.4.3 Wheel Unit on the Rim with Tyre

Measured and simulated characteristic values of the radiation pattern for IFA with conventional summer big tyre are gathered in Table B.13.

		Gain horizontal in dBi			Gain vertical in dBi			Gain total in dBi		
Meas. plane	Result	max	min	avg.	max	min	avg.	max	min	avg.
xy	Sim (Skin-effect model)	-25.6	-54.0	-30.3	-24.9	-39.5	-28.1	-23.5	-39.3	-26.1
	Sim (Media model)	-27.0	-63.4	-31.6	-26.0	-43.2	-29.6	-24.6	-40.0	-27.5
	Measurement	-26.1	-49.2	-30.6	-26.3	-36.6	-30.3	-24.6	-36.3	-27.4
yz	Sim (Skin-effect model)	-18.1	-46.6	-24.4	-29.3	-62.1	-35.2	-17.8	-45.1	-24.1
	Sim (Media model)	-20.4	-42.7	-26.2	-28.4	-53.6	-34.7	-19.8	-41.1	-25.6
	Measurement	-18.5	-48.5	-25.5	-30.1	-50.6	-37.2	-18.2	-43.3	-25.2
xz	Sim (Skin-effect model)	-26.5	-57.8	-31.2	-21.1	-39.7	-26.1	-21.0	-28.7	-25.0
	Sim (Media model)	-27.5	-60.1	-32.5	-22.8	-44.1	-27.9	-22.6	-31.4	-26.6
	Measurement	-25.6	-58.0	-31.9	-20.5	-43.2	-25.7	-20.5	-39.0	-24.8

Table B.13: Characteristic measured radiation pattern values for Wheel Unit with Inverted F Antenna on rim with big conventional summer tyre (245/40 R18 Pirelli P Zero Rosso)

Measured and simulated characteristic values of the radiation pattern for loop with conventional summer big tyre are gathered in Table B.14.

		Gain horizontal in dBi			Gain vertical in dBi			Gain total in dBi		
Meas. plane	Result	max	min	avg.	max	min	avg.	max	min	avg.
xy	Sim (Skin-effect model)	-24.2	-53.4	-27.8	-29.4	-54.1	-33.9	-24.2	-37.3	-26.9
	Sim (Media model)	-23.7	-57.5	-27.6	-28.6	-54.3	-32.7	-23.3	-35.4	-26.4
	Measurement	-23.7	-51.3	-27.5	-26.4	-45.6	-30.8	-22.6	-31.8	-25.8
yz	Sim (Skin-effect model)	-33.3	-63.1	-38.9	-18.0	-35.2	-21.9	-18.0	-35.2	-21.8
	Sim (Media model)	-37.9	-75.9	-44.6	-16.0	-34.3	-20.2	-16.0	-34.3	-20.1
	Measurement	-35.7	-69.0	-41.3	-17.3	-32.8	-20.6	-17.3	-32.5	-20.6
xz	Sim (Skin-effect model)	-17.9	-36.9	-21.6	-43.3	-62.9	-47.0	-17.9	-36.7	-21.6
	Sim (Media model)	-16.1	-40.7	-19.7	-37.2	-57.1	-42.5	-16.1	-37.3	-19.7
	Measurement	-17.0	-39.0	-20.4	-33.6	-62.6	-40.0	-17.0	-33.1	-20.4

Table B.14: Characteristic measured radiation pattern values for Wheel Unit with loop antenna on rim with big conventional summer tyre (245/40 R18 Pirelli P Zero Rosso)

Measured and simulated characteristic values of the radiation pattern for IFA with SSR summer big tyre are gathered in Table B.15. Measured and simulated characteristic values of the radiation pattern for loop with SSR summer big tyre are gathered inTable B.16.

		Gain horizontal in dBi			Gain vertical in dBi			Gain total in dBi		
Meas. plane	Result	max	min	avg.	max	min	avg.	max	min	avg.
xy	Sim (Skin-effect model)	-26.3	-64.5	-33.7	-25.0	-46.9	-31.4	-24.8	-42.7	-29.4
	Sim (Media model)	-29.4	-55.3	-37.4	-27.6	-62.9	-33.7	-27.4	-44.7	-32.1
	Measurement	-30.9	-49.3	-35.9	-27.3	-47.9	-32.1	-25.7	-39.7	-30.6
yz	Sim (Skin-effect model)	-28.2	-56.0	-35.4	-20.2	-50.0	-26.9	-19.6	-47.6	-26.3
	Sim (Media model)	-31.5	-56.0	-39.2	-23.5	-58.3	-29.8	-22.9	-51.0	-29.3
	Measurement	-28.0	-47.5	-34.7	-26.9	-54.7	-30.2	-24.7	-46.3	-28.8
xz	Sim (Skin-effect model)	-22.6	-45.1	-29.1	-25.9	-53.2	-32.4	-22.4	-41.5	-27.4
	Sim (Media model)	-25.3	-52.1	-31.9	-28.0	-74.2	-34.9	-25.2	-43.2	-30.2
	Measurement	-26.5	-36.3	-29.7	-27.0	-57.7	-32.9	-25.1	-33.0	-28.0

Table B.15: Characteristic measured radiation pattern values for Wheel Unit with Inverted F Antenna on rim with big Self Supporting Runflat summer tyre (245/40 R18 Bridgestone Potenza REO40)

		Gain horizontal in dBi			Gain vertical in dBi			Gain total in dBi		
Meas. plane	Result	max	min	avg.	max	min	avg.	max	min	avg.
xy	Sim (Skin-effect model)	-23.3	-55.3	-29.2	-30.8	-66.8	-35.5	-23.3	-36.1	-28.3
	Sim (Media model)	-26.8	-69.9	-32.6	-31.4	-61.3	-36.4	-26.7	-37.2	-31.1
	Measurement	-29.0	-55.9	-31.7	-28.7	-61.0	-33.3	-26.5	-35.7	-29.5
yz	Sim (Skin-effect model)	-39.3	-58.4	-46.0	-17.5	-41.9	-24.1	-17.5	-40.8	-24.0
	Sim (Media model)	-38.7	-72.8	-46.4	-20.0	-43.0	-26.0	-20.0	-41.5	-25.9
	Measurement	-41.9	-64.0	-47.8	-21.3	-36.3	-26.5	-21.3	-36.2	-26.5
xz	Sim (Skin-effect model)	-17.2	-48.7	-23.8	-41.7	-87.7	-46.2	-17.2	-42.4	-23.8
	Sim (Media model)	-19.6	-54.0	-26.3	-37.7	-63.7	-42.5	-19.6	-38.6	-26.2
	Measurement	-20.9	-31.8	-24.8	-34.4	-65.0	-40.8	-20.9	-31.6	-24.7

Table B.16: Characteristic measured radiation pattern values for Wheel Unit with loop antenna on rim with big Self Supporting Runflat summer tyre (245/40 R18 Bridgestone Potenza REO40)

C Optimal Positioning of Antennas within Vehicle Structure

Annex C describes the developed procedure for optimal positioning of receiver antenna within the vehicle body based on the analysis of the radiated field strength from the transmitter unit. Firstly it is explained which electromagnetic simulations were conducted. The achieved results are evaluated and visualized.

Optimal position for an antenna within a vehicle structure may be derived upon practice and experimental values. Due to necessity of conducting numerous measurements such approach is very time consuming and is still burdened with arguable grade of uncertainty. It is advantageous to support valuable experience with computational methods (Chapter 6) in order to determine the optimum antenna position.

The optimal receiver position for RKE and TPMS means, the position in which the RF receiver is receiving the possibly highest power and it's functionality is assured for all possible positions of the WU mounted on each wheel.

The underneath presented method is especially eligible if an antenna has to operate for and satisfy more than one service (Chapter 5.2). For RKE performance, any place in the vehicle is optimal, where nearly omni-directional radiation around the vehicle is achievable. This applies to any vehicle independent on the vehicle body structure. For TPMS system, the situation is far more complicated. It is difficult to answer the question about the optimal position that would satisfy best receiving performance for all four WUs simultaneously taking into account all wheel positions. The result is highly vehicle body dependent and the position has to be a weighted trade-off in order to built a good RF link between all antennas. Therefore, specific simulation approach was developed.

The first step is the electromagnetic simulation (conducted with simulation software FEKO). Simulation approach is based not on the performance examination of the receiver in various locations within the vehicle body. The strategy of the developed procedure is to activate one by one all the transmitters (WUs in this specific case) and record the resulting Electrical Field Values E within the vehicle structure. The second step is the analytical evaluation of the achieved results (conducted with numerical computation and visualization software MATLAB).

C.1 Electromagnetic Simulation Procedure

Simulation model of the vehicle under investigation without the ground is presented in Chapter 6, Fig. 6.2. The model comprises the vehicle metallic body, wheels (rim and tyre) as well as WUs. The ground during the simulations was represented by PEC plane.

The Electric Field Strength E (components E_x, E_y, E_z and E_t) was calculated within in Fig. C.2 depicted volume comprising the vehicle (Fig. C.2 FIRST STEP). Neighbouring discrete points were distributed with five centimetres distance. Each wheel during the simulation process was rotated over a full turn (with step of five degrees). Four such simulations had to take place - one for each wheel.

The above described simulations resulted in 288 values (4 wheels, for each wheel 72 WU simulation positions over a full wheel turn) for each single discrete point (Fig. C.2 SECOND STEP). With such a high result number a good evaluation procedure and algorithm was necessary.

C.2 Result Evaluation

In order to evaluate the calculated values an algorithm that would depict and compare all field values for all wheels had to be developed.

The evaluation is done separately for all field components of the Electric Field $\vec{E}$. Obtained values and geometrical vehicle interior representation (discrete points) form a 3D matrix.

The field values are sorted in descending order (Fig. C.2 THIRD STEP, obtained values matrix). The corresponding coordinates of the discrete points are written in a stack.

Parallel another 3D matrix based on vehicle interior coordinates is formed. Initially all values of this matrix are set to zero (Fig. C.3 FOURTH STEP, coordinate base matrix).

In the following step the obtained values matrix and coordinate base matrix are correlated (Fig. C.3 FIFTH STEP). The rows of obtained values matrix are evaluated one by one starting with the row including the highest electric field values. Every element of such a row has got a belonging coordinate (stack). This coordinate is found in the coordinate base matrix and in the according field the value is increased by one. There are so many correlation steps as many obtained values matrix rows or as any of the fields in coordinate base matrix reaches the value of 288. Fields with highest reached values represent the best positions for placing the receiver.

There are many coordinate base matrix fields with high values that neighbour directly to the fields with very low values. The evaluation was conducted for the densely distributed discrete points every five centimetre. The receiver biggest diameter amounted twenty centimetres. For this reason mean values for all neighbouring points within twenty centimetre distance were calculated (the vehicle interior was clustered in volumes). In this way achieved highest values represent the most optimal possible positions for the receiver. Evaluated candidate positions are depicted in Fig. C.1.

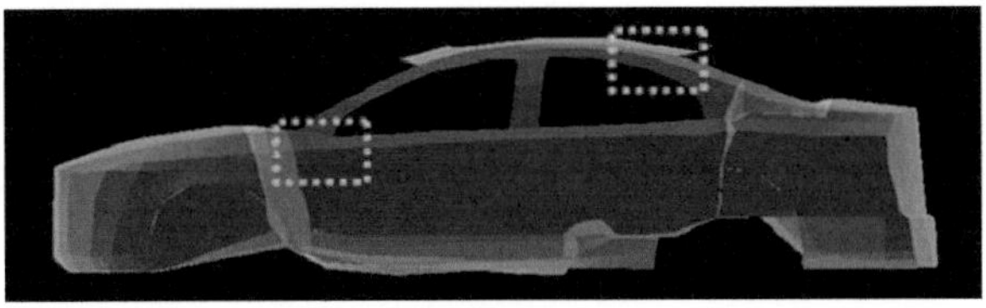

(a) Lateral vehicle view

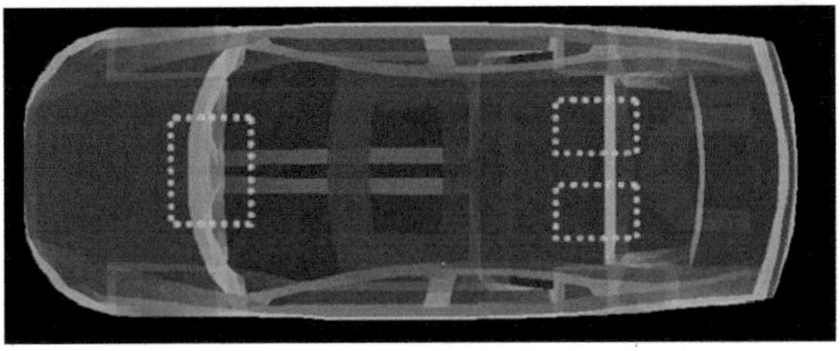

(b) Top vehicle view

Figure C.1: Optimal areas suitable for receiver placement

C.3 Result Visualization

Additionally to the developed search algorithm, a visualization tool for field distribution within the vehicle body was developed. It is possible to visualize the obtained and clustered field values as in Fig. C.4

Such visualization helps to confirm the proper functioning of the developed algorithm. Additional tool helps to view the electrical field distribution for each wheel and field component separately with the developed visualization tool. There is possibility to visualize the field distribution for each wheel position as well as an average over full wheel turn. Fig. C.5 represents an example of the visualization tool for horizontal vehicle cross section. The tool allows manual observation of the simulated field values.

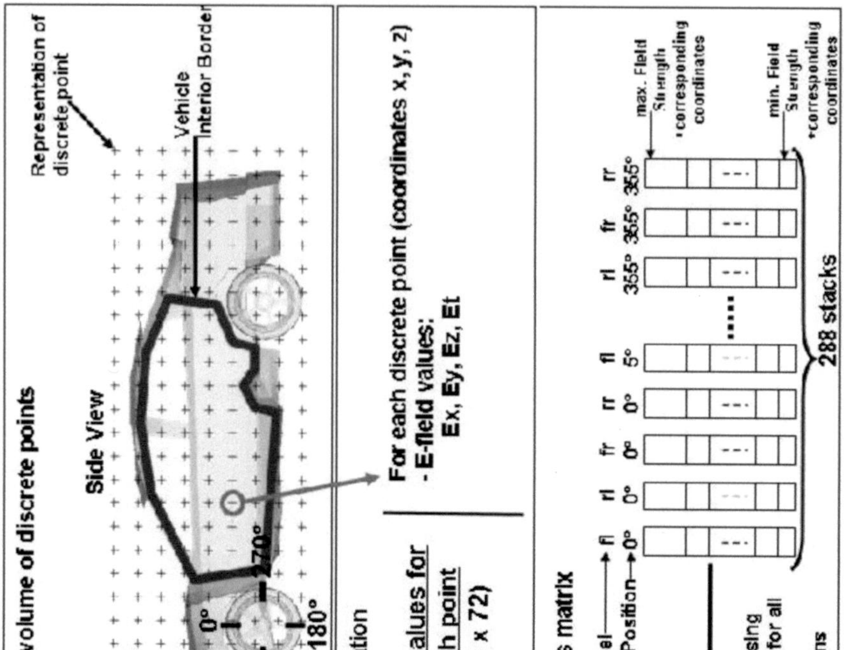

Figure C.2: Simulation and evaluation procedure for optimal antenna position within the vehicle structure

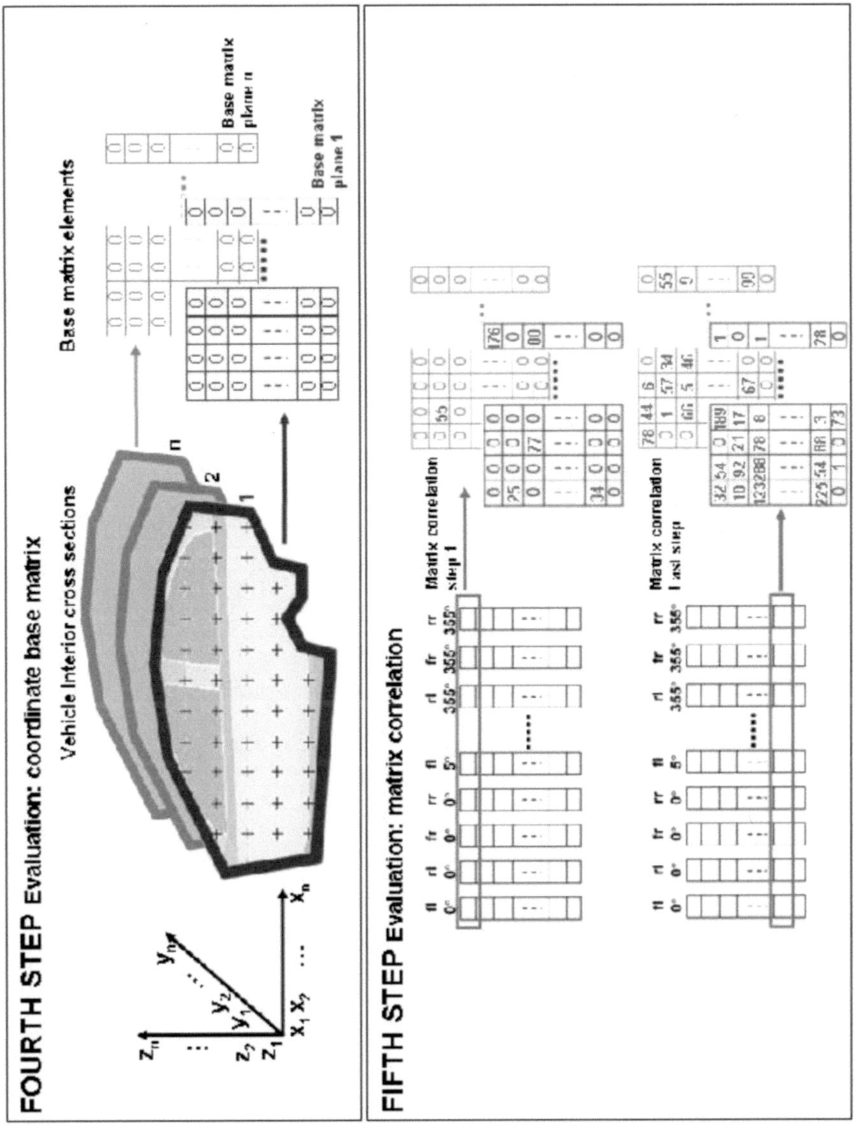

Figure C.3: Simulation and evaluation procedure for optimal antenna position within the vehicle structure

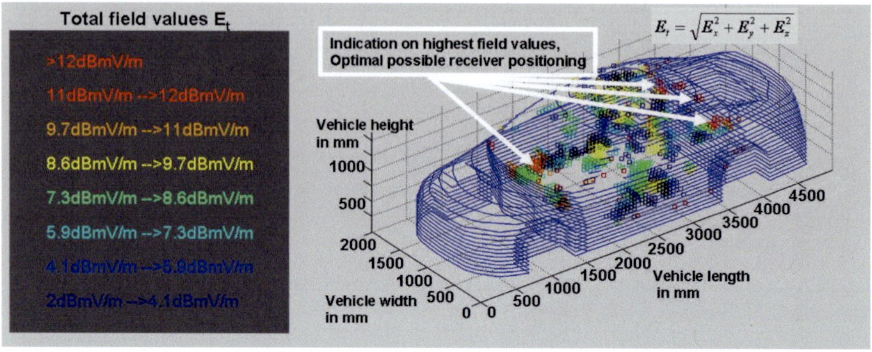

Figure C.4: Visualization of the simulated total electric field values within the vehicle interior

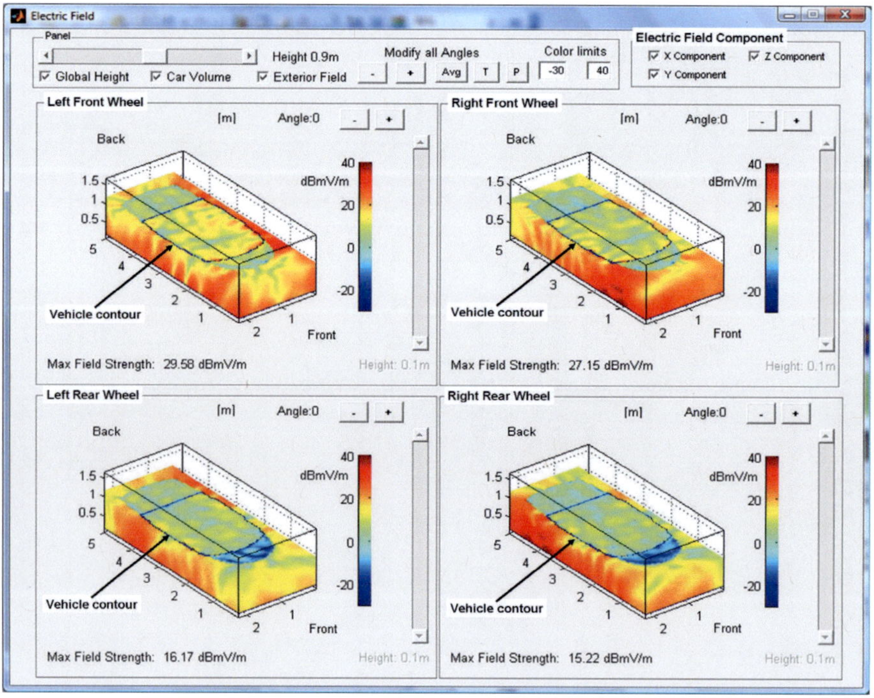

Figure C.5: Visualization tool for field distribution within the vehicle structure

D Tyre Rubber Material Measurement

Appendix D comprehends material measurement results for tyre rubber. The beneath presented results are complement to Chapter 4. Results of detailed material measurement for numerous conventional tyre samples from different tyre sections are presented. The analysis of these results allows afterwards conclusions.

D.1 Conventional Tyre Representative at Room Temperature

Conventional tyre (Pirelli P Zero Rosso) tread was measured in 5 regions (Fig. 4.11a). Measurement results of the samples resulting from these regions are depicted in Fig D.1 to Fig D.9. For each region three consequent samples were measured. The graphics show calculated average (solid line) as well as variance of the measured results (vertical solid lines). Firstly tread tyre rubber was measured together with the metallic reinforcement, second measurement for each probe was done with pure tread rubber layer (measurement without metallic reinforcement). This was the first tyre under investigation and so many measurements were made in order to assure repeatability.

Fig. D.1 and Fig. D.2 show the results for sample from region 1.
Fig. D.3 and Fig. D.4 show the results for sample from region 2.
Fig. D.5 and Fig. D.6 show the results for sample from region 3.
Fig. D.7 and Fig. D.8 show the results for sample from region 4.
Fig. D.9 and Fig. D.10 show the results for sample from region 5.

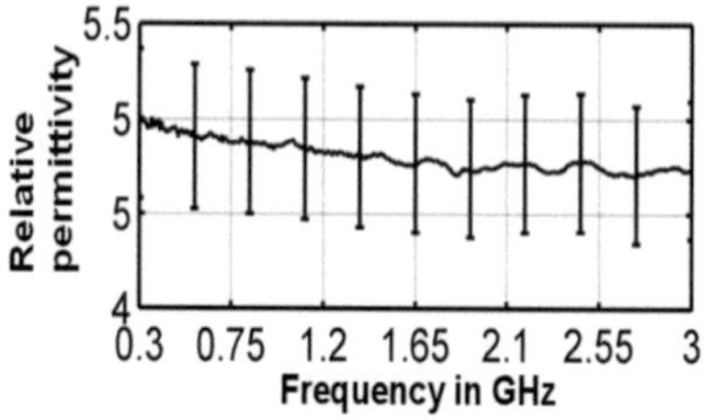

(a) Relative permittivity $\epsilon_r(f)$

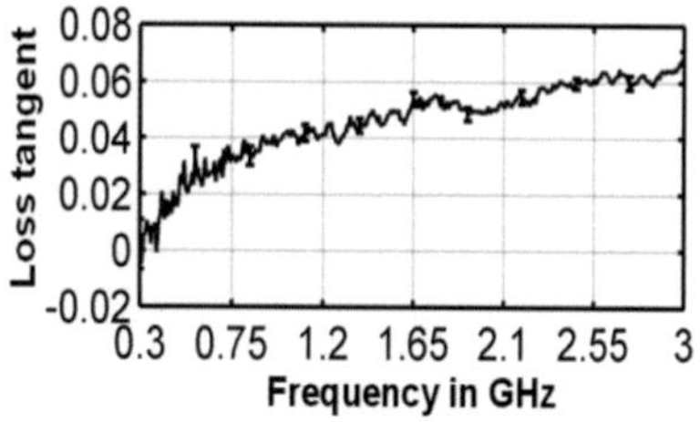

(b) Loss tangent $tan\delta(f)$

Figure D.1: Measured electrical parameters of conventional tyre, sample from region 1, with metallic reinforcement, room temperature

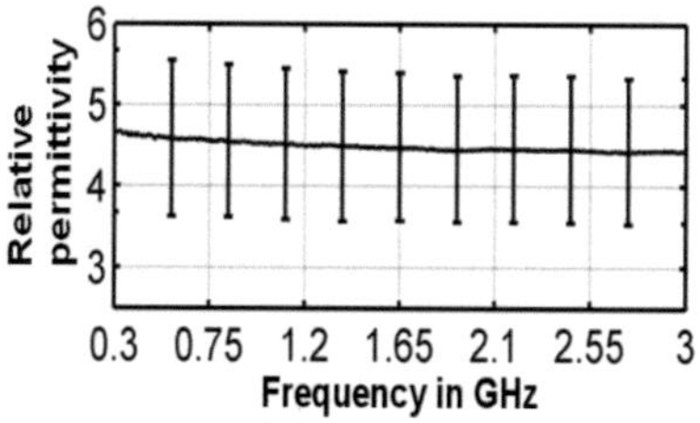

(a) Relative permittivity $\epsilon_r(f)$

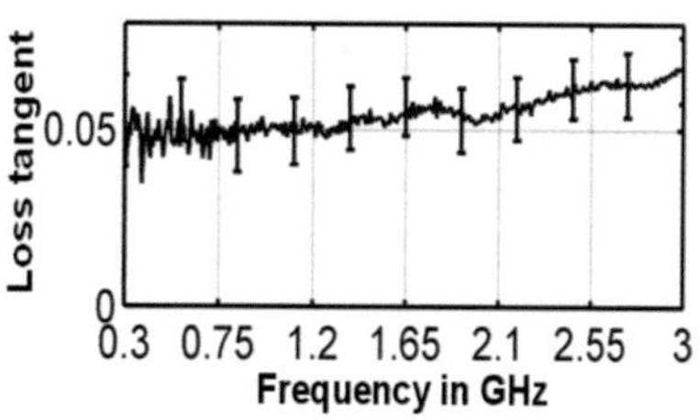

(b) Loss tangent $tan\delta(f)$

Figure D.2: Measured electrical parameters of conventional tyre, sample from region 1, without metallic reinforcement, room temperature

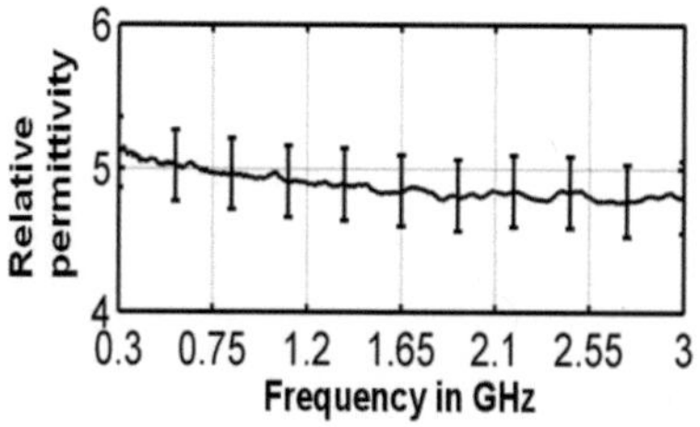

(a) Relative permittivity $\epsilon_r(f)$

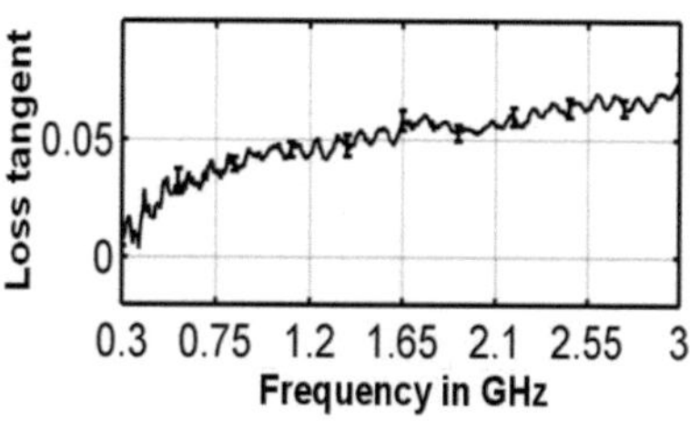

(b) Loss tangent $tan\delta(f)$

Figure D.3: Measured electrical parameters of conventional tyre, sample from region 2, with metallic reinforcement, room temperature

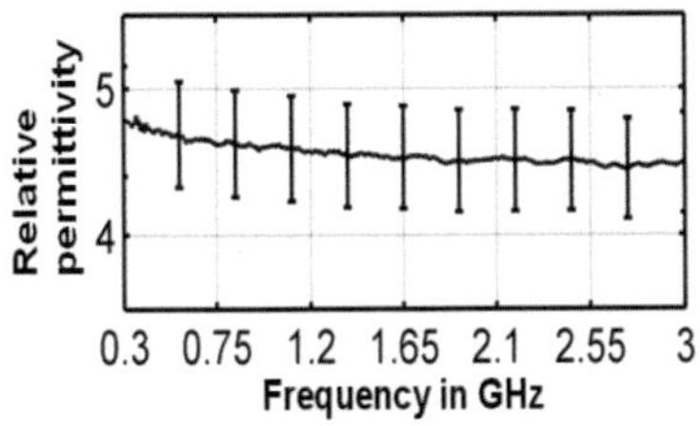

(a) Relative permittivity $\epsilon_r(f)$

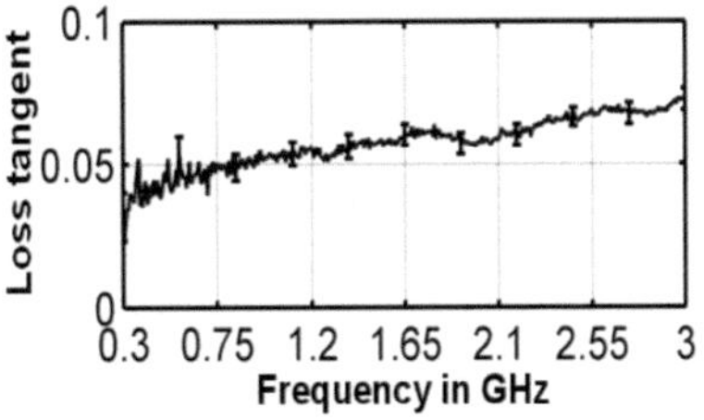

(b) Loss tangent $tan\delta(f)$

Figure D.4: Measured electrical parameters of conventional tyre, sample from region 2, without metallic reinforcement, room temperature

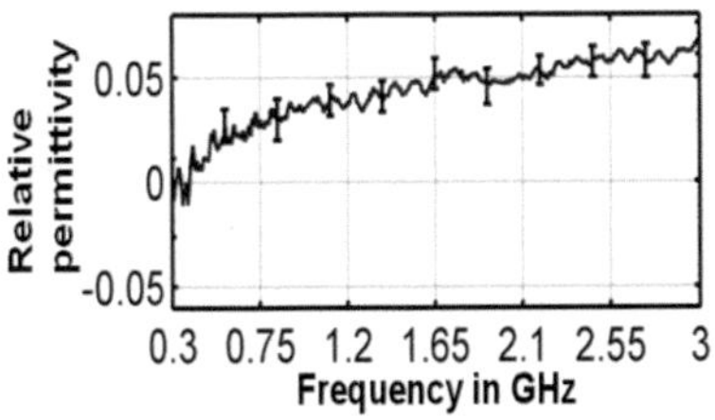

(a) Relative permittivity $\epsilon_r(f)$

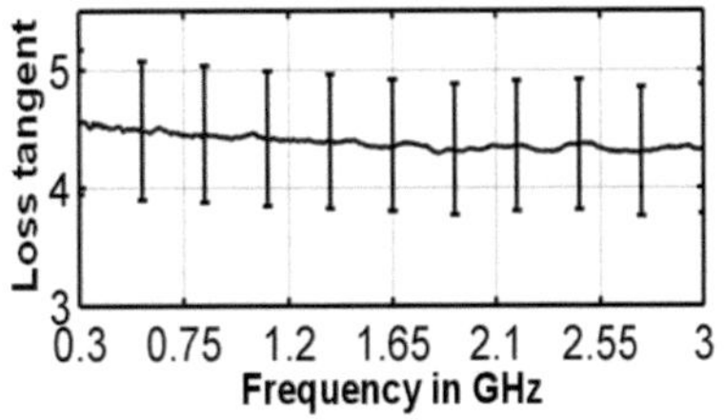

(b) Loss tangent $tan\delta(f)$

Figure D.5: Measured electrical parameters of conventional tyre, sample from region 1, with metallic reinforcement, room temperature

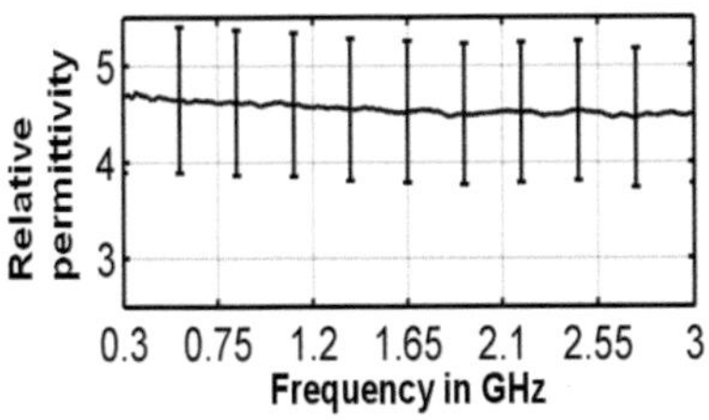

(a) Relative permittivity $\epsilon_r(f)$

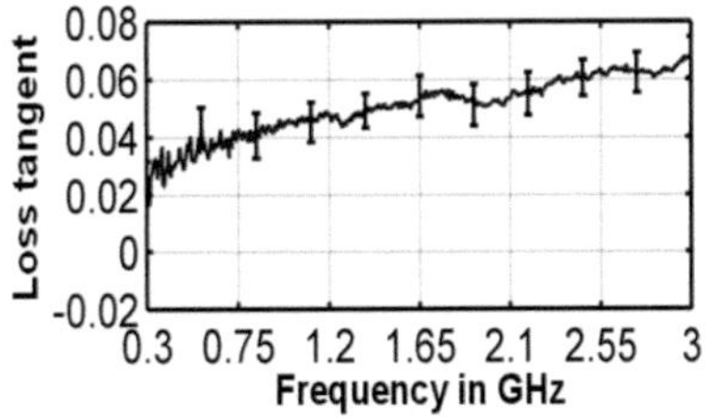

(b) Loss tangent $tan\delta(f)$

Figure D.6: Measured electrical parameters of conventional tyre, sample from region 3, without metallic reinforcement, room temperature

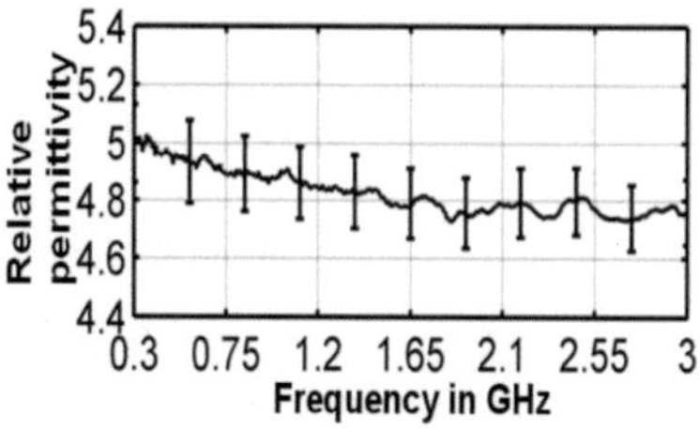

(a) Relative permittivity $\epsilon_r(f)$

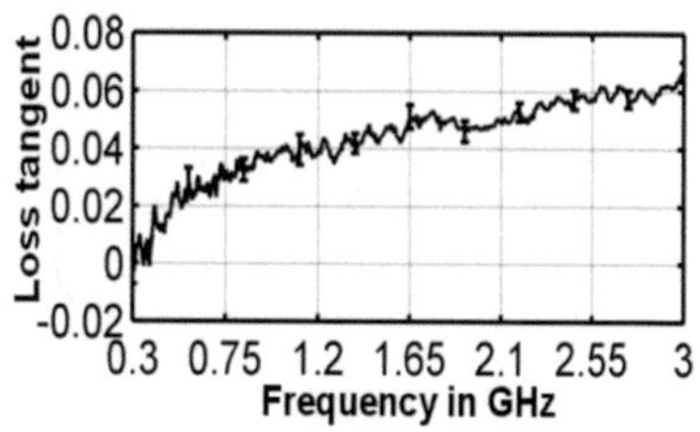

(b) Loss tangent $tan\delta(f)$

Figure D.7: Measured electrical parameters of conventional tyre, sample from region 4, with metallic reinforcement, room temperature

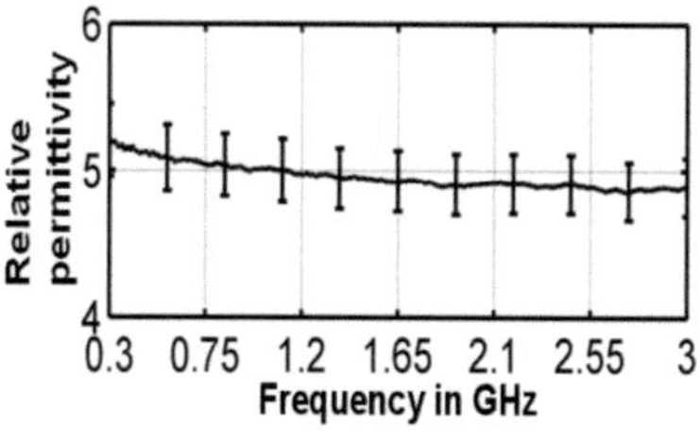

(a) Relative permittivity $\epsilon_r(f)$

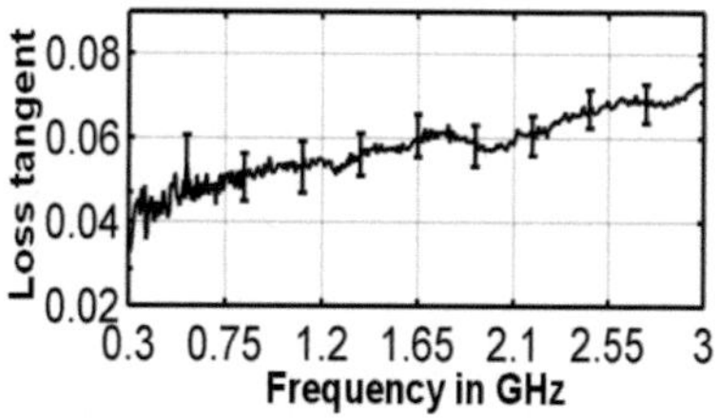

(b) Loss tangent $tan\delta(f)$

Figure D.8: Measured electrical parameters of conventional tyre, sample from region 4, without metallic reinforcement, room temperature

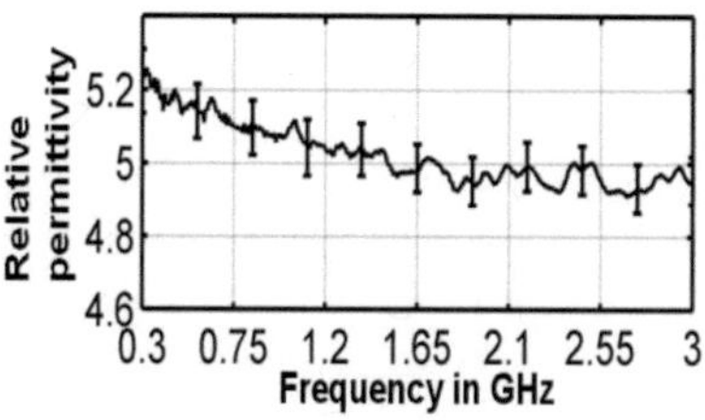

(a) Relative permittivity $\epsilon_r(f)$

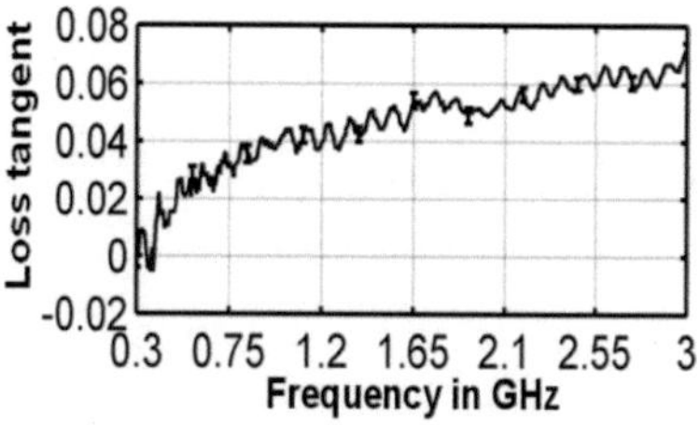

(b) Loss tangent $tan\delta(f)$

Figure D.9: Measured electrical parameters of conventional tyre, sample from region 1, with metallic reinforcement, room temperature

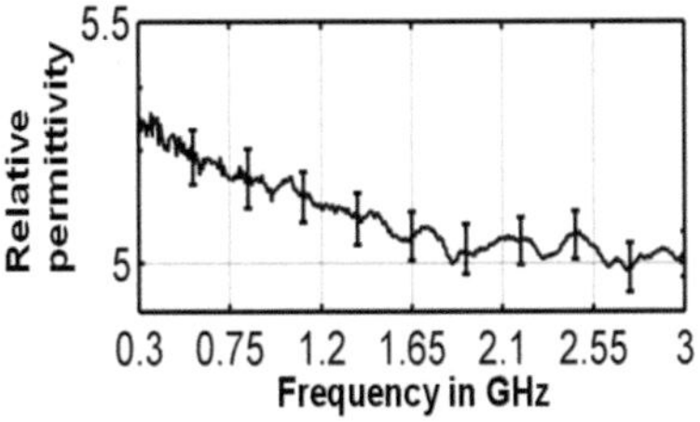

(a) Relative permittivity $\epsilon_r(f)$

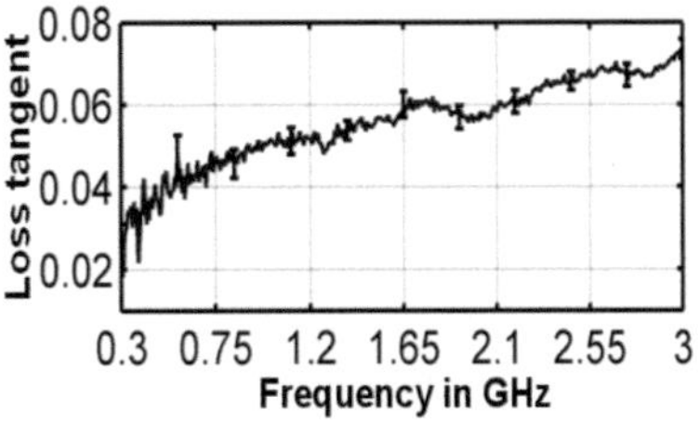

(b) Loss tangent $tan\delta(f)$

Figure D.10: Measured electrical parameters of conventional tyre, sample from region 5, without metallic reinforcement, room temperature

Alaysing the results it may be stated that:

- All measurements feature values of the same range: for relative permittivity between 3.8 and 5.3. for loss tangent in the range of 0.02 to 0.08 dependent on the frequency range.
- For the samples where the probe was consisting of pure rubber the measured values of relative permittivity are slightly smaller than for the samples with rubber and metallic reinforcement.
- For the samples where the probe was consisting of pure rubber the measured values of loss tangent are slightly higher than for the samples with rubber and metallic reinforcement.
- Overall similar dielectric properties are observed for all tyre regions. Thus tyre tread may be considered as homogeneous, independent from tread profile.

Average values of the above measurements are presented in Chapter 4.5.

With the above presented results the certainty of the chosen measurement method with dielectric probe and measurement repeatability are confirmed.

Forschungsberichte aus dem Institut für Höchstfrequenztechnik und Elektronik (IHE) der Universität Karlsruhe (TH) (ISSN 0942-2935)

Herausgeber: Prof. Dr.-Ing. Dr. h.c. Dr.-Ing. E.h. mult. Werner Wiesbeck

Forschungsberichte aus dem
Institut für Höchstfrequenztechnik und Elektronik (IHE)
der Universität Karlsruhe (TH) (ISSN 0942-2935)

Band 12 Friedhelm Rostan
Dual polarisierte Microstrip-Patch-Arrays für zukünftige satellitengestützte SAR-Systeme (1996)

Band 13 Markus Demmler
Vektorkorrigiertes Großsignal-Meßsystem zur nichtlinearen Charakterisierung von Mikrowellentransistoren (1996)

Band 14 Andreas Froese
Elektrochemisches Phasengrenzverhalten von Supraleitern (1996)

Band 15 Jürgen v. Hagen
Wide Band Electromagnetic Aperture Coupling to a Cavity: An Integral Representation Based Model (1997)

Band 16 Ralf Pötzschke
Nanostrukturierung von Festkörperflächen durch elektrochemische Metallphasenbildung (1998)

Band 17 Jean Parlebas
Numerische Berechnung mehrlagiger dualer planarer Antennen mit koplanarer Speisung (1998)

Band 18 Frank Demmerle
Bikonische Antenne mit mehrmodiger Anregung für den räumlichen Mehrfachzugriff (SDMA) (1998)

Band 19 Eckard Steiger
Modellierung der Ausbreitung in extrakorporalen Therapien eingesetzter Ultraschallimpulse hoher Intensität (1998)

Band 20 Frederik Küchen
Auf Wellenausbreitungsmodellen basierende Planung terrestrischer COFDM-Gleichwellennetze für den mobilen Empfang (1998)

Band 21 Klaus Schmitt
Dreidimensionale, interferometrische Radarverfahren im Nahbereich und ihre meßtechnische Verifikation (1998)

Band 22 Frederik Küchen, Torsten C. Becker, Werner Wiesbeck
Grundlagen und Anwendungen von Planungswerkzeugen für den digitalen terrestrischen Rundfunk (1999)

Band 23 Thomas Zwick
Die Modellierung von richtungsaufgelösten Mehrwegegebäudefunkkanälen durch markierte Poisson-Prozesse (2000)

Forschungsberichte aus dem
Institut für Höchstfrequenztechnik und Elektronik (IHE)
der Universität Karlsruhe (TH) (ISSN 0942-2935)

Band 24 Dirk Didascalou
Ray-Optical Wave Propagation Modelling in Arbitrarily Shaped Tunnels (2000)

Band 25 Hans Rudolf
Increase of Information by Polarimetric Radar Systems (2000)

Band 26 Martin Döttling
Strahlenoptisches Wellenausbreitungsmodell und Systemstudien für den Satellitenmobilfunk (2000)

Band 27 Jens Haala
Analyse von Mikrowellenheizprozessen mittels selbstkonsistenter finiter Integrationsverfahren (2000)

Band 28 Eberhard Gschwendtner
Breitbandige Multifunktionsantennen für den konformen Einbau in Kraftfahrzeuge (2001)

Band 29 Dietmar Löffler
Breitbandige, zylinderkonforme Streifenleitungsantennen für den Einsatz in Kommunikation und Sensorik (2001)

Band 30 Xuemin Huang
Automatic Cell Planning for Mobile Network Design: Optimization Models and Algorithms (2001)

Band 31 Martin Fritzsche
Anwendung von Verfahren der Mustererkennung zur Detektion von Landminen mit Georadaren (2001)

Band 32 Siegfried Ginter
Selbstkonsistente Modellierung der Erhitzung von biologischem Gewebe durch hochintensiven Ultraschall (2002)

Band 33 Young Jin Park
Applications of Photonic Bandgap Structures with Arbitrary Surface Impedance to Luneburg Lenses for Automotive Radar (2002)

Band 34 Alexander Herschlein
Entwicklung numerischer Verfahren zur Feldberechnung konformer Antennen auf Oberflächen höherer Ordnung (2002)

Band 35 Ralph Schertlen
Mikrowellenprozessierung nanotechnologischer Strukturen am Beispiel von Zeolithen (2002)

Forschungsberichte aus dem
Institut für Höchstfrequenztechnik und Elektronik (IHE)
der Universität Karlsruhe (TH) (ISSN 0942-2935)

Band 36 Jürgen von Hagen
Numerical Algorithms for the Solution of Linear Systems of Equations Arising in Computational Electromagnetics (2002)

Band 37 Ying Zhang
Artificial Perfect Magnetic Conductor and its Application to Antennas (2003)

Band 38 Thomas M. Schäfer
Experimentelle und simulative Analyse der Funkwellenausbreitung in Kliniken (2003)

Band 39 Christian Fischer
Multistatisches Radar zur Lokalisierung von Objekten im Boden (2003)

Band 40 Yan C. Venot
Entwicklung und Integration eines Nahbereichsradarsensorsystems bei 76,5 GHz (2004)

Band 41 Christian Waldschmidt
Systemtheoretische und experimentelle Charakterisierung integrierbarer Antennenarrays (2004)

Band 42 Marwan Younis
Digital Beam-Forming for high Resolution Wide Swath Real and Synthetic Aperture Radar (2004)

Band 43 Jürgen Maurer
Strahlenoptisches Kanalmodell für die Fahrzeug-Fahrzeug-Funkkommunikation (2005)

Band 44 Florian Pivit
Multiband-Aperturantennen für Basisstationsanwendungen in rekonfigurierbaren Mobilfunksystemen (2005)

Band 45 Sergey Sevskiy
Multidirektionale logarithmisch-periodische Indoor-Basisstationsantennen (2006)

Band 46 Martin Fritz
Entwurf einer breitbandigen Leistungsendstufe für den Mobilfunk in Low Temperature Cofired Ceramic (2006)

Band 47 Christiane Kuhnert
Systemanalyse von Mehrantennen-Frontends (MIMO) (2006)

Band 48 Marco Liebler
Modellierung der dynamischen Wechselwirkungen von hoch-intensiven Ultraschallfeldern mit Kavitationsblasen (2006)

Forschungsberichte aus dem Institut für Höchstfrequenztechnik und Elektronik (IHE) der Universität Karlsruhe (TH) (ISSN 0942-2935)

Band 49 Thomas Dreyer
Systemmodellierung piezoelektrischer Sender zur Erzeugung hochintensiver Ultraschallimpulse für die medizinische Therapie (2006)

Band 50 Stephan Schulteis
Integration von Mehrantennensystemen in kleine mobile Geräte für multimediale Anwendungen (2007)

Band 51 Werner Sörgel
Charakterisierung von Antennen für die Ultra-Wideband-Technik (2007)

Band 52 Reiner Lenz
Hochpräzise, kalibrierte Transponder und Bodenempfänger für satellitengestützte SAR-Missionen (2007)

Band 53 Christoph Schwörer
Monolithisch integrierte HEMT-basierende Frequenzvervielfacher und Mischer oberhalb 100 GHz (2008)

Band 54 Karin Schuler
Intelligente Antennensysteme für Kraftfahrzeug-Nahbereichs-Radar-Sensorik (2007)

Band 55 Christian Römer
Slotted waveguide structures in phased array antennas (2008)

Fortführung als
"Karlsruher Forschungsberichte aus dem Institut für Hochfrequenztechnik und Elektronik" bei KIT Scientific Publishing
(ISSN 1868-4696)

Karlsruher Forschungsberichte aus dem Institut für Hochfrequenztechnik und Elektronik (ISSN 1868-4696)

Herausgeber: Prof. Dr.-Ing. Thomas Zwick

Die Bände sind unter www.ksp.kit.edu als PDF frei verfügbar oder als Druckausgabe bestellbar.

Band 55 Sandra Knörzer
Funkkanalmodellierung für OFDM-Kommunikationssysteme bei Hochgeschwindigkeitszügen (2009)
ISBN 978-3-86644-361-7

Band 56 Thomas Fügen
Richtungsaufgelöste Kanalmodellierung und Systemstudien für Mehrantennensysteme in urbanen Gebieten (2009)
ISBN 978-3-86644-420-1

Band 57 Elena Pancera
Strategies for Time Domain Characterization of UWB Components and Systems (2009)
ISBN 978-3-86644-417-1

Band 58 Jens Timmermann
Systemanalyse und Optimierung der Ultrabreitband-Übertragung (2010)
ISBN 978-3-86644-460-7

Band 59 Juan Pontes
Analysis and Design of Multiple Element Antennas for Urban Communication (2010)
ISBN 978-3-86644-513-0

Band 60 Andreas Lambrecht
True-Time-Delay Beamforming für ultrabreitbandige Systeme hoher Leistung (2010)
ISBN 978-3-86644-522-2

Band 61 Grzegorz Adamiuk
Methoden zur Realisierung von dual-orthogonal, linear polarisierten Antennen für die UWB-Technik (2010)
ISBN 978-3-86644-573-4

Band 62 Jutta Kühn
AlGaN/GaN-HEMT Power Amplifiers with Optimized Power-Added Efficiency for X-Band Applications (2011)
ISBN 978-3-86644-615-1

Karlsruher Forschungsberichte aus dem Institut für Hochfrequenztechnik und Elektronik (ISSN 1868-4696)

Band 63 Małgorzata Janson
Hybride Funkkanalmodellierung für ultrabreitbandige MIMO-Systeme (2011)
ISBN 978-3-86644-639-7

Band 64 Mario Pauli
Dekontaminierung verseuchter Böden durch Mikrowellenheizung (2011)
ISBN 978-3-86644-696-0

Band 65 Thorsten Kayser
Feldtheoretische Modellierung der Materialprozessierung mit Mikrowellen im Durchlaufbetrieb (2011)
ISBN 978-3-86644-719-6

Band 66 Christian Andreas Sturm
Gemeinsame Realisierung von Radar-Sensorik und Funkkommunikation mit OFDM-Signalen (2012)
ISBN 978-3-86644-879-7

Band 67 Huaming Wu
Motion Compensation for Near-Range Synthetic Aperture Radar Applications (2012)
ISBN 978-3-86644-906-0

Band 68 Friederike Brendel
Millimeter-Wave Radio-over-Fiber Links based on Mode-Locked Laser Diodes (2013)
ISBN 978-3-86644-986-2

Band 69 Lars Reichardt
Methodik für den Entwurf von kapazitätsoptimierten Mehrantennensystemen am Fahrzeug (2013)
ISBN 978-3-7315-0047-6

Band 70 Stefan Beer
Methoden und Techniken zur Integration von 122 GHz Antennen in miniaturisierte Radarsensoren (2013)
ISBN 978-3-7315-0051-3

Band 71 Łukasz Zwirełło
Realization Limits of Impulse-Radio UWB Indoor Localization Systems (2013)
ISBN 978-3-7315-0114-5

Band 72 Xuyang Li
Body Matched Antennas for Microwave Medical Applications (2014)
ISBN 978-3-7315-0147-3

Karlsruher Forschungsberichte aus dem Institut für Hochfrequenztechnik und Elektronik (ISSN 1868-4696)

Band 73 Sebastian Diebold
Transistor- und Leitungsmodellierung zum Entwurf von monolithisch integrierten Leistungsverstärkern für den hohen Millimeterwellen-Frequenzbereich (2014)
ISBN 978-3-7315-0161-9

Band 74 Christian Rusch
Integrierte, planare Leckwellenantennen für 3D-Millimeterwellen-Radarsysteme basierend auf dem holografischen Prinzip (2014)
ISBN 978-3-7315-0234-0

Band 75 Marlene Harter
Dreidimensional bildgebendes Radarsystem mit digitaler Strahlformung für industrielle Anwendungen (2014)
ISBN 978-3-7315-0249-4

Band 76 Michael A. Baldauf
Abhängigkeit der Exposition von der Zellgröße beim Mobilfunk unter Gewährleistung der Versorgung (2015)
ISBN 978-3-7315-0308-8

Band 77 Alicja Ossowska
Highly Resolved Synthetic Aperture Radar with Beam Steering (2015)
ISBN 978-3-7315-0315-6

Band 78 Małgorzata Dominika Brzeska
RF Modelling and Characterization of Tyre Pressure Sensors and Vehicle Access Systems (2015)
ISBN 978-3-7315-0348-4